THE BORDER MULTIPLE

Border Regions Series

Series Editor: Doris Wastl-Walter, University of Bern, Switzerland

In recent years, borders have taken on an immense significance. Throughout the world they have shifted, been constructed and dismantled, and become physical barriers between socio-political ideologies. They may separate societies with very different cultures, histories, national identities or economic power, or divide people of the same ethnic or cultural identity.

As manifestations of some of the world's key political, economic, societal and cultural issues, borders and border regions have received much academic attention over the past decade. This valuable series publishes high quality research monographs and edited comparative volumes that deal with all aspects of border regions, both empirically and theoretically. It will appeal to scholars interested in border regions and geopolitical issues across the whole range of social sciences.

The Border Multiple

The Practicing of Borders between Public Policy and Everyday Life in a Re-scaling Europe

Edited by

DORTE JAGETIĆ ANDERSEN
University of Southern Denmark

MARTIN KLATT
University of Southern Denmark

MARIE SANDBERG
University of Copenhagen, Denmark

Routledge
Taylor & Francis Group
LONDON AND NEW YORK

First published 2012 by Ashgate Publishing

2 Park Square, Milton Park, Abingdon, Oxon OX14 4RN
711 Third Avenue, New York, NY 10017, USA

Routledge is an imprint of the Taylor & Francis Group, an informa business

First issued in paperback 2016

British Library Cataloguing in Publication Data
The border multiple : the practicing of borders between public policy
and everyday life in a re-scaling Europe. — (Border regions series)
1. Border security—Europe. 2. Border security—Government policy—Europe.
3. Border crossing—Europe. 4. Territory, National—Europe. 5. Europe—Boundaries.
I. Series II. Andersen, Dorte Jagetić. III. Klatt, Martin,
1964– IV. Sandberg, Marie.
320.1'2'094–dc23

Library of Congress Cataloging-in-Publication Data
The border multiple : the practicing of borders between public policy and everyday life in a re-scaling Europe / [edited by] Dorte Jagetić Andersen,Martin Klatt and Marie Sandberg.
p. cm. — (Border regions series)
Includes bibliographical references and index.
ISBN 978-1-4094-3708-6 (hardback : alk. paper) — ISBN 978-1-4094-3709-3 (ebook)
1. Border security—Europe. 2. Borderlands—Europe. 3. Border crossing—Europe.
4. Europe—Emigration and immigration—Government policy.
I. Andersen, Dorte, 1971– II. Klatt, Martin, 1964– III. Sandberg, Marie.
JV7590.B67 2012
320.1'2—dc23

2012004876

ISBN 13: 978-1-4094-3708-6 (hbk)
ISBN 13: 978-1-138-26987-3 (pbk)

Contents

List of Figures

List of Tables

List of Contributors

Dorte Jagetić Andersen is Associate Professor in the Department of Border Region Studies, University of Southern Denmark.

Marta Byrska-Szklarczyk is a Graduate of the Institute of Ethnology and Cultural Anthropology at the University of Warsaw, Poland.

Anna Gawlewicz is a Ph.D. fellow in the School of Geography, University of Leeds, UK.

Jouni Häkli is Professor of Regional Studies at the University of Tampere, Finland.

Devan Jagodic is Assistant Professor at the Slovene Research Institute, Trieste, Italy.

Michael Keating is Professor of Politics at the University of Aberdeen School of Social Science, Scotland, and chair of Scottish Politics.

Martin Klatt is Associate Professor in the Department of Border Region Studies, University of Southern Denmark.

Olivier Thomas Kramsch is Senior Lecturer in the Department of Human Geography, Radboud University, Nijmegen, The Netherlands.

Martin Lemberg-Pedersen, Ph.D., is an External Lecturer at the Section of Philosophy, the Department of Media, Cognition and Communication, University of Copenhagen, Denmark.

Gergő Medve-Bálint is a Junior Research Fellow at the Institute for Political Science, Center for Social Sciences at the Hungarian Academy of Sciences, and a Ph.D. Fellow at Central European University, Budapest, Hungary.

Marie Sandberg is Assistant Professor in Ethnology at the Saxo Institute, University of Copenhagen, Denmark.

Sara Svensson is a Ph.D. fellow at the Central European University, Budapest, Hungary.

Carsten Yndigegn is Associate Professor in the Department of Border Region Studies, University of Southern Denmark.

Thomas M. Wilson is Professor of Anthropology at Binghamton University, State University of New York, US.

Acknowledgements

This book is the result of the conference 'A Borderless Europe?' arranged in a collaboration between Department for Border Regions Studies, University of Southern Denmark, and the Saxo Institute, Ethnology Section, University of Copenhagen, held in Sonderborg, September/October 2010. Together with the invited keynote speakers, around 60 border scholars from all over Europe contributed to a lively atmosphere and a range of high quality workshop presentations during the 3 days of the conference. As indicated in the question mark of the conference title, discussions and presentations provided evidence and gave rich illustrations of the fact that despite discourses of a 'borderless world', the borders of contemporary Europe still have an impact on everyday life practices within and outside of the realms of the EU. The editors would like to thank all those who contributed to the discussions during the conference. A special thanks to Doris Wastl-Walter for her contribution to the conference as well as her comments on the book's contents in the early stages of its development. In editing this volume, we are particularly grateful for the meticulous work done by Hans-Kristian Chapman Pedersen, who had the task of preparing the book for camera-ready submission. We also thank all authors who contributed to the volume, not the least for their timeliness in getting the contributions to us, including Olivier Kramsch who willingly has contributed to this volume with an Afterword. Finally, thanks go to both CEMES, Centre for Modern European Studies at the University of Copenhagen and the Department of Border Region Studies, University of Southern Denmark for covering the expenses of editing the book.

Introduction

Dorte Jagetić Andersen and Marie Sandberg

In recent years, European borders have changed character. We have witnessed an opening of borders in the Schengen area, where national borders seem less important than they were previously, something, which creates new opportunities for mobility across borders. The status of EU internal borders stems from the Schengen Agreement, designed to serve the free movement of persons within the territory it applies to and secure the territory by means of harmonized visa, asylum, and migration policies and a broad cooperation between police, immigration and judicial authorities (Fairhurst, 2010). Interaction across the open Schengen borders increases and people get used to crossing borders without the traditional passport control. When Denmark announced the reintroduction of regular customs control at her borders with Germany and Sweden in June 2011, this resulted in a public outcry especially in Germany, including calls for boycott of Denmark as a holiday destination by German politicians. Apparently, free travel is treasured by many Europeans even if open borders awaken anxiety with regard to security.

Political cooperation across borders between regional and local actors is also a common sight at today's inner and outer EU borders. Since the 1990s, the EU has generously supported cross-border projects with the Interreg Program, today Objective Three of the Cohesion Policy. On the external borders, PHARE, TACIS and the European Neighbourhood Policy (ENP) have included similar policies. Some researchers see cross-border regions evolving as political entrepreneurs in a system of multilevel governance (Perkmann, 2007; Schmitt-Egner, 2001; Scott, 2000). However, several case studies reveal limits to cross-border cooperation and that such cooperation remains an elite project borne by a limited number of often highly motivated and committed agents within very weak institutional frameworks, while the broad public tends to be indifferent (Barten et al., 2006; Bucken-Knapp, 2001; Dupeyron, 2008; Ehlers, 2001; Harguindéguy and Bray, 2009; Klatt, 2011; Liikanen, 2008; Strüver, 2004; Strüver, 2005).

While Euroregions have initiated many cross-border activities, they have not been able to significantly reduce the technical and legal barriers still present at the inner Schengen borders. Also their legal status has remained weak, as it cannot be compared with the legal status of public authorities within a country. Euroregions do not have decisive competencies, and their governance status can be characterized as soft governance only, if at all. Although hardly physically visible, the EU borders within the Schengen zone have in many respects conserved their dividing impact. Cross-border labour markets are, for instance, hindered by different regulations, lack of information, as well as cultural and linguistic differences. So there is a continuing dilemma as to how to make the last step forward to establish factual cross-border regions to overcome the still divisive character of the inner Schengen borders.

As a parallel process, and due to a fear of rising international crime rates, problematic immigration patterns as well as the fight against terrorism, the European Union's security regime has tightened control over its territorial borders and at the same time extended its regulatory practices into neighbouring countries such as Frontex-policing and the intensification of joint actions against cross-border crime. Hence, European integration has gained some Central-Eastern European borders the status of the new external borders of Europe. Consequently, they have experienced increased attention in border literature within the last decade, giving a remarkable insight into the EU and Schengen extensions and most notably – and in line with academic and public discourse – stressing the isolating effect on some EU neighbouring countries. Although the media and politicians often focus on the rhetoric of bridge-building of an internal borderless Europe, the academic research into external EU borders reports many cases of bordering-othering, social, political, and economic exclusion and, not the least, the closure of the EU.

Apart from its risk-reducing function, the Schengen agreement acts as an intelligent tool letting in those who bring wealth (or valuable personal, educational, business, skills, etc.) whilst at the same time keeping out those who bring burden (Kramsch et al., 2004). This results in the emergence of an exclusive zone difficult to enter for those who for various reasons do not fit the profile. The ENP was introduced to tighten the cooperation of the member states at the fringes of the EU with the neighbouring non-EU members (Boedeltje et al., 2006; Orbie, 2008). ENP aims at increasing prosperity by means of extending the internal market and establishing preferential trade relations, including relatively easy access for sought-after legal inhabitants on one side and tackling cross-border threats on the other (Commission of the European Communities, 2003). Its introduction implies further incentives such as prevention of organized crime and other security threats, development of strong democratic institutions and respect for human rights, cultural cooperation, research cooperation, educational exchange, and integration of communications infrastructure. European neighbourhood policy thus becomes a form of Europeanization of the EU neighbouring states by use of soft politics.

Hence, the often used symbol of 'Fortress Europe' (Klein, 2003; Kofman and Sales, 1992) sounds too far-fetched to describe the present situation at the EU external borders. Van Houtum and Pijpers suggest that instead we should consider the EU as 'a gated community' (2007). The closure analogy, they argue, does not apply to the clever selection policy introduced by the European Union. The gated community analogy, on the other hand, expresses the whole complex nature of the supranational strive for protection and fear-caused isolation. Other researchers use the symbol of 'a Euro-curtain', which divides the European continent, or a computer 'firewall' as a symbolic figure for filtering-processes (Bialasiewicz and O'Loughlin, 2002; Mrinska, 2006; Szmagalska-Follis, 2009).

The Scope of this Book

Apparently a distinction exists between the regulatory practices performed internally to the Schengen area and those performed externally; the Schengen agreement changes the character of European borders and it leaves populations living in the border regions located externally to the EU, and particularly in the East of Europe, with different life conditions than those they had only a few years ago. However, rather than looking at these de- and re-bordering processes as constitutive of a geographical dichotomy between a borderless inside and an excluded outside, this anthology addresses the processes from the perspectives of a) a more general re-scaling of the European territorial landscape b) the multiplicity of border practices in which borders are performed and c) the mobility turn and corresponding relational approaches to everyday life and public policy. By using a set of illustrative examples from everyday life border practices and dynamics, as well as cross-border cooperation, the distinction between inside and outside become less apparent; internal regulatory practices have not, as anticipated, completely disappeared, and even when these might sometimes be of a slightly different nature, challenges and obstacles for mobility across borders are just as apparent in the internal area as they are when crossing external borders. To state that regulatory practices are significantly more present at the external Schengen borders than they are internally to the Schengen area would thus be to overlook both the continuity of European border regimes and the complexity of bordering processes when performed in practise. We use the verb 'performed' deliberately in order to stress how borders are not pre-given but rather to be seen as effects of the practices through which they are made (Butler, 1993; Mitchell, 2006; Mol, 2002).

Studying Borders in a Rescaling Europe

It is our impression that in order to analyse the complexity of bordering processes in contemporary Europe it is necessary to move beyond the internal–external dichotomy and instead connect these processes with more general economic and political changes occurring not only internally to Europe but at a global scale (van Houtum et al., 2005; Rumford, 2006; Pavlakovich-Kochi et al., 2004). The insight that European borders have changed character in recent years is in other words not simply a consequence of the introduction of the Schengen agreement. The changes as well as the introduction of the Schengen agreement relates to more general processes in accordance with which Europe's territorial landscape is currently re-scaling:

> The important point for our purposes is that globalization, European integration and the wider processes of economic change did not suppress territory as such but have in important ways disassociated it from the nation state. We are seeing

> not so much the end of territory as its re-invention and reshaping. (Keating, Chapter 1 this volume)

Although the extent can be discussed, globalization-processes clearly reduce the ability of states to control their territories, be it economically, politically, or culturally. State authorities are pressured from above as well as within, witnessing the development of a new competitive regionalism encouraging territorial concentration in terms of economic processes, political participation as well as cultural identification (Keating, 1998, 2003). There is now a vast amount of literature emphasizing the importance of local and regional territory for economic prosperity and distribution as well as how spatial developments depend less on hard factors, like infrastructure and subsidies and more on soft factors such as human and social capital, knowledge and innovation – people skills in other words (cf. for instance Scott, 1996).

The various processes of change create a variety of new functional, social, political and cultural spaces, which may or may not coincide (Keating in this volume). The extent to which they do so is not the result of functional restructuring performed at EU level but of political agency, that is, the strategies of a variety of actors creating new institutional forms and endowing them with capacity at various institutional levels. In this way, the European territorial landscape is being shaped by different forces, be they local or regional, supra-national or international, societal or cultural, and the territorial landscape become more complex than ever before.

New Security Regimes

These changes in territorialization processes in Europe as well as their complexity reflect in the security regimes supposed to guard European borders from unwanted entrances. As illustrated in Martin Lemberg-Pedersen's article in this anthology, the European border controls have disseminated into diverse practices informed not only by recent globalisation processes but more importantly by an attempt to limit such processes through a reformulation of colonial geopolitics as well as new networks attempting to secure Europe against the (supposedly) otherwise unlimited flow of unwanted migrants. These processes have created a new colonial European frontier but unlike the traditional colonial border which existed in geographical space because of their enforcement by European territorial regimes, the new European frontier is geographically dispersed and instituted by means of various regulatory practices. As Lemberg-Pedersen puts it:

> A strict focus on legal decisionism bypasses how the European geopolitics of mobility is realised via a wide array of practices such as administrative deportations, Frontex-policing and financial programmes. The sovereign power becomes an underdetermined abstractum [...] the current EU-regime is transforming the notion of European sovereignty by invoking new configurations

> of borders and space, the power to control these and protectionary loopholes. (Lemberg-Pedersen, Chapter 2 this volume)

Again this shows how the security regimes, according to which European borders institute, cannot be conceptualised by way of an internal/external distinction existing in concrete geographical space. The complexity of the processes as well as their sometimes abstract, sometimes (very) concrete nature, infer that the security regime is spatially present everywhere and that it might institute in places where we would least expect it. In order to conceptualize European borders in the contemporary context it is thus necessary to develop new approaches, methods, and conceptual tools that are able to capture the multiplicity of practices through which borders are performed as well as the way in which inside and outside are defined differently in time and space (see for example Andersen, Chapter 7, Häkli, Chapter 8 and Sandberg, Chapter 6, herein).

Multiplying Scale

In the light of this new understanding of territory as a rescaling of Europe and the new security regimes leads us to point at the need within border studies to recuperate our understanding of space and territory and ask what the status of borders and border regions are (van Houtum et al., 2005). Are we in other words witnessing the same regulatory practices when we visit the Ukrainian–Polish border as we are on the Hungarian–Croatian? And is cross-border cooperation the same matter whether you are a local politician in the Basque country or in the Ruhr distinct? In regards of internal Schengen borders, it is first of all necessary to multiply the notion of scale and try to conceptualize as well as visualize the absent non-border in its functional inter-relations. Internal borders matter – be they imagined or be they practically significant – and this despite of processes removing them from our visual landscape. As scholars of border studies it is our task to make the invisible border appear, so to speak. With regards to external Schengen borders, we have to understand how the East of Europe has become a neutral buffer territory between global powers and, accordingly re-conceptualize how older archetypes are revitalized as lines and limits of former imperial powers (Maier, 2006). Finally, we need to re-conceptualize the continued dynamics of inclusion and exclusion at the outer borders of Europe.

The book approaches the re-scaling of Europe with the hypothesis that current re- and de-bordering processes in Europe are in a state of transition, or even crisis and, as argued by Keating, that we are not witnessing the end of territory but it's re-scaling. Regulatory practices and the image of borders, consequently, involve a multiplicity of new border practices which are first of all not easily grasped and secondly combating each other. Hence, instead of simply portraying the internal borders as territorial borders becoming more open and the external as territorial borders becoming more solid, the book shows how European borders generate and are generated by a range of different practices and mental images, including cross-border mobility and cooperation, biopolitics, neo-colonialism, spatial re-orderings and networking, as well as reiterations of traditional border practices. European borders of today should be understood in the light of diverse possibilities of practice opened in their redefinition.

The Practice Turn in Border Studies

This book gives an insight into how we can conceptualize and grasp the mobile and changing character of the European landscape, and how we are able to empirically investigate border practices in the light of these processes of change. Following our assumptions, the book approaches borders from the micro-level drawing more or less implicitly on Chris Rumford's concept of borderwork (Rumford, 2008). The concept of borderwork has developed due to the need to both sharpen and broaden our understanding of borders and the changed significance they hold in contemporary society. In Rumford's outlook borders have become floating corresponding to the flows of networks, and take on diverse forms. They can be characterized as permeable and closed at the same time, both framing state sovereignty and allowing unhindered flow. Like a computer firewall, they perform intelligent filtering of immigrants, being open for the attractive and closed for the unwanted (see also Walters, 2006a, 2006b).

Borders are thereby conceptualized as practices that are situated and constituted in the specificity of political negotiations as well as in the everyday life performance of them. To investigate and understand these practices that can be both specific and diverse, the book integrates functional and symbolic assets to the construction of borders. It examines how people who are directly affected by the presence/emergence of borders or their disappearance try to adapt to and even interact with these complex processes of de- and re-bordering. In de- and re-bordering processes borders are being territorially displaced and border controls are, in principle, being carried out by anyone anywhere. It is the local inhabitants that call the police when they spot an illegal trespasser, it is the all-encompassing surveillance, and it is the flight companies that perform passport control for the immigration authorities at the site of departure (Rumford, 2008). The book emphasizes this ability people have to participate in the making of borders as well as the empowerment that can result from border activities. Focus is hereby extended from border practices of the state to the practices of ordinary people and the way in which they construct, shift, and dismantle borders (Galasiński and Meinhof, 2002; Meinhof and Galasiński, 2002; Meinhof, 2002; Schultz et al., 2002; Löfgren, 2008; Sandberg, 2009a, 2009b; Linde-Laursen, 2010).

The Border Multiple

To understand the complexity of European borders it is necessary to avoid essentializing current trends and instead think in terms of heteronomy and diversity; what we refer to as 'The Border Multiple'. Choosing this particular title for the book, we pay tribute to the Dutch philosopher Annemarie Mol, who in the book *The Body Multiple* (2002) scrutinizes the treatment of the disease atherosclerosis at a Dutch hospital. She thereby provides a precise and useful definition of 'multiplicity' as well as its constitution in practice. Multiplicity refers not simply to diversity but points to the fact that the different ways any given object or phenomenon is handled also enact specific versions of it. In the case of the treatment of atherosclerosis the vein is for example being cut into with a scalpel, bombarded with ultrasound as well as put under a microscope. Everyone who partakes in this process of treatment; the surgeon, the doctor, the laboratory technician, enact a version of the disease. Mol conceptualizes this set of practices as different versions of the same object and characterizes it as 'the multiple object' (ibid.). The different practices enact slightly different versions of the disease, yet they are all handling the same disease and together they make up a multiple reality.

Hence, we conceptualize borders as performed or enacted effects of the multiple practices through which they are made. Consequently, while working through conceptual schemas, such as differentiating between processes of de-and re-bordering, or hard versus soft borders, we must not reify them nor insist that there is only one way of apprehending our new spatial reality. We must think creatively about what borders are in this European spatial landscape, something, which foremost involves showing for whom they are and how those for whom they are, enact and perform them. Otherwise, we repeat a political agenda currently reinforcing the age-old dichotomy between the East and the West, where internal borders are understood as harmonious and conceptualized as best practices, whereas external borders are seen as much more precarious and – exactly therefore – in need of strict control.

Practicing Borders and History

The emphasis within border studies on the making of borders within everyday life practices inscribes itself into a wider practice turn within contemporary social theory. The turn implies, as according to Schatzki, a shared understanding of practices as 'embodied, materially mediated arrays of human activity' (2001, p. 2). This field of practice oriented approaches is diverse and includes also internal disputes, especially as to how and by what sorts of entities these activities are (inter)mediated, and whether non-human actors can be regarded as entities 'having' practices or not (see Häkli, Chapter 8 and Sandberg, Chapter 6, herein). What unites scholars of practice theory, however, is 'a shared interest in establishing the field of practices as the place to study the nature and transformation of their subject matter' (Schatzki, 2001, p. 2).

This characterization of a common ground for scholars of practice theory can also be applied to the diverse field of approaches within border studies. As we argue in this book the orientation towards practice and the multiple ways borders are enacted in everyday life provide a shared point of departure for border scholars.

For instance, when borders and the impacts of borders are analysed in practice, it becomes very evident that 'borders are still very much with us', despite discourses of globalization proclaiming the existence of a borderless world. Adding to this quest for practice oriented border studies is a – maybe growing – need to incorporate historical dimensions into these practicing of European borders:

> [T]he consideration of space and groups of people in border studies must be tempered with a consideration of their histories. History and space intersect the framing, construction and practices of power and the politics of identity and culture in borderlands. Moreover, the historical reflexivity which inspires so much of the critique of the national state and its borders must also be applied to other geopolitical borders, such as those of regions and localities. While these latter sorts of borders owe a great deal to both the real and ideal of state borders, together with state and now supranational borders in Europe they give meaning and substance to the transformations that seem to vitalize so much of meaning in the daily lives of Europeans. (Wilson, Chapter 4, herein)

A practice oriented border study in other words needs to be conceptualized through dimensions of history in order to show how borders are constituted through specific intersections between geography and history. The border multiple is in its own peculiar ways an offspring of these intersections.

We thus propose to work from the perspective of what Benedict Anderson has called 'The Inverted Telescope' (1998), asking ourselves how we can use micro-scale border practices to both conceptualize and visualize what European borders are at a more general level. By focusing on border practice and by using methods developed within fields as diverse as human geography, sociology, ethnology, and historical research, it is possible to illuminate the particular, and also very different, significance people ascribe to borders and how it depend on their interactions with it. It is our belief that through the integration of this perspective into mainstream understandings of geopolitical and functional assets of borders, we can broaden our conception of borders in relation to ongoing processes of de- and re-bordering in Europe.

Borders and the Mobility Turn

EU border region policies and initiatives have, as previously mentioned, aimed at increasing mobility internally to the Schengen area. Free movement as a fundamental right for EU citizens is inscribed in the free movement laws of the Treaty of Rome 1957 and later the Lisbon strategy (Favell, 2009, p. 177f). The freedom of movement as a founding principle of the EU is further described in the FEU Treaty (treaty of the Functioning of the EU) in the form of the free movement of goods, citizenship of the Union, the free movement of workers, the freedom of establishment, the free movement of services and the free movement of capital

(European Commission, 2011). These four freedoms, and with a newly added fifth, the free movement of knowledge, provides the European citizen with the ability to travel, settle, work, and study freely across the intra-European borders of EU27, and there are no visas required. Incorporated within these mobility policies of the EU was the assumption, or hope even, that the more we move ourselves across borders the more we integrate.

Yet, as Favell's analyses of differential spectres of the EU migratory systems show: the 'free movers' of European Union space need to be seen as one particular and rather small group of intra-European migrants, whom he designates the 'Eurostars' (Favell, 2008a). And further, in fact rather than increased integration the reverse tendency seems to be at stake: increased mobility leads to less integration (Favell, 2009, p. 176). The crossing or transgression of, or the mobility across borders does not necessarily lead to the dissolving or the negation of borders; quite to the contrary: often the very functioning of borders become more evident when they are transgressed (cf. O'Dowd 2010).

The more central placing of borders in the study of society is also due to an increasing interest in networks, flows, and mobilities (Canzler et al., 2008; Larsen et al., 2006; Urry, 2008). This interest has been characterized as 'The new Mobilities Paradigm' (Sheller and Urry, 2006) covering a multidisciplinary field of scholars from anthropology, STS, Geography, Cultural Studies, Tourism and Migration Studies. As opposed to classical conceptions of travelling within transport studies the new mobility paradigm wishes to show the complex relations between places and people. Further the mobility approach emphasizes that movement and immobility, as well as travelling and dwelling, are always strongly interconnected: 'The new paradigm emphasizes how all mobilities entail specific often highly embedded and immobile infrastructures' (ibid., p. 210). Practices of mobility are for example most likely accompanied by practices of 'homing' or regrounding new places which also implies that they are located and materialized. Illustrative examples of how EU mobility policies effect complex mobility patterns can for instance be discerned in practices of student mobility (as in Gawlewicz and Yndigegn, Chapter 9, herein) and of residential mobility across borders (as in Jagodic, Chapter 10, herein). A third area encompassing mobility across borders is cross-border cooperation within Europe which is discussed and exemplified in the last part of this book.

European Cross-border Cooperation

Studies of European cross-border cooperation are also an integral part of the border multiple. Traditionally, cross-border cooperation was a matter of foreign policy, regulated in treaties between sovereign states. This changed after World War II, when more and more local authorities in border regions claimed the authority to enter cooperation agreements with local authorities across the border. In 1980, the Council of Europe passed the Outline Convention on Transfrontier Cooperation, which for the first time opened for binding cooperation agreements between regional and local authorities across a border. Most EU member states have since concluded treaties with their neighbours allowing the establishment of institutionalized local and regional cross-border cooperation. Since the 1990's, the EU has supported such cooperation financially through the Interreg program establishing cross-border cooperation across virtually all European borders.

Public policy approaches in border studies have, in line with EU visions, concentrated their investigations of border regions on those networks of agents, which are deliberately engaged in dismantling border, that is, different kinds of cross-border partnerships and cooperation (Perkmann, 2007; Scott, 2010; Scott, 2006; Boedeltje et al., 2006). A Euroregion usually consists of a number of regional and local authorities as members, who have agreed to increase cross-border cooperation and set up a formal institution to govern this cooperation. A permanent secretariat serves the Euroregion, and the members have established a representative assembly to discuss cross-border issues and decide on cross-border activities. Some Euroregions are also responsible for managing the Interreg projects within their region. In 2007 a new legal framework for cross-border territorial cooperation was introduced with the European Grouping on Territorial Cooperation (EGTC). This, for the first time, provided a cross-border legal framework for cooperation in EU law.

To understand the significance of cross-border cooperation there is a need to empirically investigate the ways in which this cooperation disseminates into diverse form at local and regional level. Much cross-border cooperation today might be an effect of regionalization processes governed by EU policies but they are not simply conducted in accordance with these policies and implemented by bureaucrats and other professionals. More importantly, cross-border cooperation articulates independently of EU policies at the level of the individual region or locality, and sometimes in the opposite direction of the EU's intention (as illustrated by Svensson and Medve-Bálint as well as Klatt, Chapters 11 and 3, herein). Moreover, much cooperation happens for reasons other than those of EU policy making, thus also involving different means and ends than those initially intended by the EU. Hence, our studies of cross-border cooperation cannot rely on the normative framework of EU initiatives dictating what we are allowed to take into consideration in our investigations. Further, and as argued by Häkli, international boundaries tend to prove pervasive 'even in contexts where the freedom of mobility across boundaries has been a defining feature of the development of the borderland over the past decades' (Häkli, Chapter 8, herein). This is also to say that we need to move beyond the presumption that border regions are simply just laboratories of European integration and instead understand them as bringing their own multiple dynamics into the European political landscape (cf. Scott, 2005).

Examining the policy dimensions to everyday lives lived at, with, and maybe across the European borders therefore forms our last and final call with this book. The issue of how policies of different kind – be that policies outgoing from EU, state, regional, or municipal level – are sought implemented or (dis)impact the local life of international borderlands are often included in border studies.

However, often the relations between policy making and everyday life remain rather un-conceptualized, as argued by Andersen in this volume. In approaching and scrutinizing the effects and impacts of different ways of policing the European borders, Wilson argues for an 'anthropology of policy':

> I suggest that ethnographers must continue to consider what anthropologists *used to do* (at least in greater numbers and in more situations) more generally in their practical, applied and political anthropologies. I make this suggestion with an appended query: can or even should anthropologists do an anthropology of policy without continual if not constant reference to bounded places, spaces and groups, and political, legal and social institutions, framed within a locality, among many other real, bounded social and political units? This compounded question is particularly relevant for anthropologists and other scholars who research Europe's borders, whether from a perspective of a borderless or a bordered world. (Wilson, Chapter 4, herein)

This again points to the fact that new approaches to the study of cross-border cooperation are strongly needed.

Structure of the Book

The first part of the book discusses how Europe is currently re-scaling and how its new security regimes materialize in a variety of new regulatory border practices. In Chapter 1, 'Rescaling Europe', Michael Keating provides us with a frame of reference in which we are able to think the territorial reconfiguration of Europe. The article first of all problematizes modernist state theory, which has for more than a century conceptualized European states as unified entities constituted by territorially defined geographical boundaries. European states have most certainly aspired to create such unified territories limited by stable borders but the degree to which they have succeeded can, on the one hand, be discussed and is, on the other hand, highly diversified. The somewhat idealized but also distorted image of territory and space associated with the Westphalian state system, the author argues, is impossible to uphold in a contemporary context as it is being challenged by pressures from globalization-processes as well as from below and within, meaning from the level of local and regional actors. In the light of these processes, the territory re-scales and diversifies, and we are today confronted with a landscape of multiple actors influencing the way in which space and territory can be perceived and conceptualised.

Martin Lemberg-Pedersen's chapter, 'Forcing Flows of Migrants: European Externalization and Border-induced Displacement', introduces us to the new security regime, which has developed in Europe due to globalization pressures, thus linking the discussion of a re-scaling Europe with issues of mobility patterns and migration control. The chapter argues that international negotiations, external governance and extra-territorialization are all aspects of externalization of the security regime and suggest that the notion of 'borderscapes' should be adopted as analytical tool for understanding the multiple abstractions of knowledge, practices and technologies at work in EU border control. Borderscapes cannot stand alone, however, but must be complemented with the more governance and process-oriented analytical perspective of 'biopolitics' (cf. for instance Balzacq, 2009) as well as a critical focus on sovereign power and black letter law bypassing political actors involved in the depoliticization and securitization of European asylum policies (cf. for instance van Munster, 2009). Together these perspectives provide a diversified and adequate picture of the European border regime.

In Chapter 3, 'Competitive Concepts of Regionalization and Cross-border Cooperation in the Danish–German Border Region', Martin Klatt focuses on different regional concepts applied in time to the space, which today forms the border region between Denmark and Germany. In the course of history, several concepts have been applied to this space to legitimize rule and create regional identity. Saint Cults from medieval times, kings' as well as dictators' and democratically elected politicians' public orchestration of their power have followed similar patterns in constructing regional space and identity. While some concepts have been short-lived, and especially cross-border concepts did not prove viable, the regional concept of Schleswig-Holstein proved sustainable from the middle ages until the present day as a successful symbiosis of national and regional identity. Klatt analyses the interplay between national, regional and local governance in the successful or non-successful social construction of regions and regional identity, thereby showing how the processes in which Europe currently re-scales display historical continuity beyond the Westphalian system. His contribution is thus an example of how the current re-scaling of European territory is a neither unique nor timeless phenomenon but part of ongoing historical processes in which scale and space are being created.

The second part of the book includes anthropological research in everyday life practice at borders, providing examples of the direction this research can take. Thomas M. Wilson's chapter, 'Revisiting the Anthropology of Policy and Borders in Europe', introduces the research paradigm in a discussion of the need for empirical, ethnographic research in border research and a criticism of border studies. With border studies involving a diversified field of scholarly approaches, Wilson poses the question: what, then, draws border scholars together? Despite the provocative tone of the question, Wilson further points to the fact that there are a range of themes shared. Taking his own discipline of social anthropology as a point of departure, Wilson exemplifies how a shared theme is manifest in contributions demonstrating how the borders of a seemingly 'borderless' Europe still have impact on the EU member states, even though the physical manifestations as state borders are often transformed. Wilson continues by discussing how, especially, the historical dimension needs continuously to be taken into account in order to avoid reductionist typologies and assertions about the uniqueness of the role of the borders in contemporary Europe. Further, Wilson examines the relations between the anthropology of borders and public policy, also relating this to the need for 'bringing history back in'. By investigating the different intersections between space and history, politics and power, the anthropology of borders, it is argued, will be more equipped to contribute in debates of the current de- and re-bordering processes of Europe.

In Chapter 5, 'Borders from the Perspective of 'Ants': Petty Smugglers from the Polish–Ukrainian Border Crossing in Medyka', Marta Byrska-Szklarczyk presents her ethnography conducted among so-called 'Ants' who earn their living as petty cigarette smugglers. The Polish Ants (living in the nearby Polish city of Przemyśl) sell cheap Ukrainian cigarettes with a profit on the Polish side of the border. Having the border as their 'workplace' these petty smugglers, the majority of which are middle-aged women, have an everyday experience with the border fundamental to the existence of this local community. Its meaning is manifested in a complex set of contexts, imageries and mythologies within the local perception and knowledge of the border including 'culture' traits such as a specific local language, rituals, perception of space, time and physicality, power relations, and economic transactions. Byrska depicts the empirically rich repertoire of metaphors for the border used by the Ants: Metaphors of life and death of the border; the border as movement; the border as a circle and a factory; the border as a meeting point and the end of the world; and the border as a prison and war front. Through these metaphors, Byrska illuminates the multiplicity of meanings inscribed in the concept of the border and how it is intrinsic to the everyday life of Ants.

In Chapter 6, 'Border Orderings: The Co-existence of Border Focusing and Letting Border Issues take the Back Seat at the German–Polish Border' Marie Sandberg departs from the turn to practice within border studies, suggesting a further opening of the floor not only for actors of everyday life practices but for a broader concept of discursive-material actors. Sandberg introduces the notion of 'modes of ordering' as an approach for grasping the multiplicity of actors contributing to the ways borders are created, shifted and transformed in everyday life. The main characteristic of the modes of ordering, originally developed by scholar of science and technology John Law, is that they describe recurring patterns in the different ways relations between discursive-material actors are organized. Sandberg explores the modes of ordering in an analysis discerning recurring patterns in the handling and organizing of German–Polish cross-border integration in the cities of Görlitz and Zgorzelec. The multiplicity of approaches is analysed as different, yet co-existing modes of ordering. Exemplifying what can be characterized 'a cross-border regime', it has been a goal for the two cities to enhance integration across the historically burdened Oder–Neisse border. Despite intentions, however, the various ways of transgressing the border do not seem to mobilize one of the main targets of integration, namely the youth. On this backdrop the analysis shows, how some modes of ordering interfere with those modes of ordering that, through the means of projects funded and supported by municipal, regional, state, and EU levels, aim at crossing and transgressing the border.

In her chapter, ‘The Multiple Politics of Borders: Doing the Croatian–Slovenian Border from the Perspective of an Ethnographer on the Move’, Dorte Jagetić Andersen argues that in order to grasp the complexity of politics related to the existence of borders, we need to study the politics of borders as reflected in everyday life border practice and not simply as geopolitical decision making or public policy making. By way of ethnographic research on the Slovenian–Croatian border on the Istrian peninsula, a context where the existence of state and imperial borders has for centuries been highly controversial, and by focusing on a diverse range of stories constructing borders, the chapter illustrates how borders are Janus-faced, contested, and contradictory at different levels of practice, be it in the realm of memory and as imagined borders, in the realm of political discourse and geopolitics, or be it in practices enacting borders in the functional realm of administration. The chapter thus emphasizes multiple borders but also the fact that to do research on mobile and multiple borders, the researcher must be mobile and possess tools that are able to move along with the research object.

The third part of the book scrutinizes how cross-border cooperation and ways in which cross-border mobility and lack thereof express at local level, and this in different ways, guided by a diverse set of actors coping and responding to the re-scaling European landscape. In the chapter, ‘Boundary Objects in Border research: Methodological Reflections with Examples from two European Borderlands’ Jouni Häkli explores a novel approach within border research of boundary objects. Using this concept as a methodological guideline Häkli analyses the complex communications and interactions of cross-border cooperation based on examples from the Catalan borderlands between France and Spain and from the Swedish–Finnish border, in particular the twin towns of Haparanda-Tornio. Based on ethnographic approaches and inspiration from Actor Network Theory (ANT) the boundary object approach stresses the entanglements of social and material realities and elaborates on the concept of ‘translation’ (developed by Bruno Latour, Michel Callon and John Law). The concept of the boundary object, originally coined by Star and Griesemer (1989) is defined as the crafting of a common reference point that can take on many shapes and that does not require consensus as end result. The concept is applied and further developed by Häkli who shows how boundary objects can function as intermediaries between different practices that are thus enabled to communicate across borders.

In Anna Gawlewicz and Carsten Yndigegn’s chapter, ‘The (In)visible Wall of Fortress Europe? Elite Migrating Youth Perceiving the Sensitive Polish–Ukrainian Border’, images and experiences that youth from Poland and Ukraine respectively have of and with the Polish–Ukrainian border are portrayed. By focusing on youths’ personal narratives of lived border experiences, the authors address the border practices of a part of the European population constituting a growing number of transnational migrants, but nevertheless border practices, which tends to be neglected in the broad body of border literature. The youth in question conceptualizes the border as both geopolitical space and a mentally developed collective construction, perceiving and negotiating the existence of the border with all its distinctive features such as everyday border checkpoint practices, transnational-scale coexistence, thus imagining and inventing 'the Other' and its EU-related significance. The study shows a clear imbalance influencing these migration patterns where Ukrainian youth choose Poland as a study destination more often and more willingly than Polish students choose Ukrainian institutions. Despite of the fact that the metaphor of 'Fortress Europe' is not directly applicable in a functional sense in the case of mobile youth, it is nevertheless reinforced by experiences and perceptions of borders.

In his chapter, 'Living (beyond) the Border – European Integration processes and Cross-border Residential Mobility in the Italian–Slovenian Border Area', Devan Jagodic sheds light on an aspect of European cross-border mobility which is so far little researched, namely cross-border residential mobility and its effects on local attachment and senses of belonging. The process of cross-border residential mobility, which can be observed in several European border regions, refers to the migration movements produced by borderlanders, who decide to buy a house and move to the other side of the border, at the same time as maintaining strong functional and affective relations with their previous country. Returning to Istria, now focusing on the Italian–Slovenian border, the article raises the question of whether or not we can understand cross-border residential mobility as the development of a new form of European transnationalism. While the residential mobility of border crossers is clearly an effect of the functional possibilities opened by European integration, a question mark has to remain as to whether this form of mobility is truly transnational. As the case illustrates, national affinities may become even stronger when people decide to live 'on the other side of the border'.

Finally, and returning to the topic of cross-border cooperation, Sara Svensson and Gergö Medve-Bálint in their chapter, 'Explaining Coverage: Why Local Governments in Central Europe do or do not join Euroregions?' answer the question; what are the local incentives to join a Euroregion as opposed to non-engagement or fostering bilateral contacts, such as village partnerships, school cooperation or civil society exchanges. They synthesize various factors suggested in the literature on Euroregions into a comprehensive theoretical framework to thereby assess the relative weight of these factors in relation to the ability of Euroregions to attract local governments' in the area where they operate. The focus is on Central Eastern Europe. While the cohesion of a cross-border region in terms of Euroregion membership may be unproblematic in many places of Western Europe, this seems to be less so in CEE which has seen many cases of overlapping or even competing initiatives. The authors suggest that in order to explain coverage, it is necessary to study a broader scope of local actors and their incentives than has been done previously in studies of cross-border cooperation and this in order to gain a better insight into the mechanisms driving cross-border cooperation at the local level.

References

Anderson, B. (1998), *The Spectre of Comparison – Nationalism, South East Asia and the World*, London: Verso.

Balzacq, T. (2009), 'Frontiers of Governance: Understanding the External Dimension of EU Justice and Home Affairs', in T. Balzacq (ed.) T*he External Dimension of EU Justice and Home Affairs: Governance, Neighbours, Security*, Basingstoke: Palgrave Macmillan, pp. 1–34.

Barten U., Bröcker J., Herrmann H., Klatt M. (2006), 'Barrieren und Potentiale der grenzüberschreitenden Zusammenarbeit in der Fehmarnbelt Region', *Beiträge aus dem Institut für Regionalforschung der Universität Kiel*, Institut für Regionalforschung der Universität Kiel, Kiel.

Berg, E. and Ehin, P. (2006), 'What Kind of Border Regime is in the Making? Towards a Differentiated and Uneven Border Strategy', *Cooperation and Conflict* 41(1): 53–71.

Bialasiewicz, L. and O'Loughlin, J. (2002), 'Galician Identities and Political Cartographies on the Polish–Ukrainian Border' in D.H. Kaplan and J. Häkli (eds) *Boundaries and Place: European Borderlands in Geographical Context*, Lanham, MD: Rowman & Littlefield, pp. 217–38.

Boedeltje, F.J., van Houtum, H. and Kramsch, O.T. (2006), 'Communication from the European Union: The European Neighbourhood Policy (ENP) "The Latest Addition to our Democratization Toolbox"', 'Eudimension' State of the Art Strategy Paper, Nijmegen.

Bucken-Knapp G. (2001), 'Just a Train-Ride Away, but Still Worlds Apart: Prospects for the Øresund Region as a Binational City', *GeoJournal* 54: 51–60.

Butler, J. (1993), *Bodies that Matter, On the Discursive Limits of 'Sex'*, New York: Routledge.

Canzler, W., Kaufmann, V. and Kesselring, S. (2008), 'Tracing Mobilities – An Introduction', in W. Canzler, V. Kaufmann and S. Kesselring 2008 (eds), *Tracing Mobilities. Contributions from the Cosmobilties Network*, Aldershot: Ashgate, pp. 11–17.

Commission of the European Communities (2003), *Wider Europe – Neighborhood: a new Framework for Relations with our Southern and Eastern Neighbours*, Communication from the Commission to the Council and the Parliament. Brussels: Commission of the European Communities.

Dupeyron B. (2008), *L'Europe au défi de ses régions transfrontalières. Expériences rhénane et pyrénéenne*, Peter Lang, Berne.

Ehlers N. (2001), 'The Utopia of the Binational City', *GeoJournal* 54: 21–32.

European Commission (2011), Single Market, General Policy Framework, available at http://ec.europa.eu/internal_market/top_layer/index_1_en.htm [accessed 12.12.2011].

Favell, A. (2008), 'The New face of East–West Migration in Europe', *Journal of Ethnic Migration Studies* 34(5): 701–16.

Favell, A. (2009), 'Immigration, Migration, and Free Movement in the Making of Europe', in J.T. Checkel and P.J. Katzenstein (eds), *European Identity*, Cambridge: Cambridge University Press, pp. 167–89.

Fairhurst, J. (2010), *Law of the European Union*, Harlow: Pearson/Longman.

Flenley, P. (2008), 'Russia and the EU: The Clash of New Neighbourhoods?', *Journal of Contemporary European Studies* 16(2): 189–202.

Galasinska, D. and Meinhof, U.H. (2002), 'Looking Across the River: German–Polish Border Communities and the Construction of the Other', *Journal of Language and Politics* 1(1): 23–58.

Harguindéguy J.-B. and Bray, Z. (2009), 'Does Cross-Border Cooperation Empower European Regions? The case of INTERREG 111-A France – Spain', *Environment & Planning C: Government & Policy* 27:747-760. DOI: 10.1068/c08119.

Keating, M. (1998), *The New Regionalism in Western European*, Aldershot: Edward Elgar.

Keating, M. (2003), 'The Invention of Regions: Political Restructuring and Territorial Government in Western Europe', in N. Brenner, B. Jessop, M. Jones and MacLeod, G. (eds), *State/Space: A Reader*, Cambridge: Blackwell, pp. 256-277.

Klatt M. (2011), 'Grenzüberschreitende politische Zusammenarbeit', in J. Beck and B. Wassenberg (eds), *Grenzüberschreitende Zusammenarbeit leben und erforschen (2): Governance in deutschen Grenzregionen. Beiträge aus dem Forschungszyklus zur grenzüberschreitenden Zusammenarbeit der Universität Strassburg und des Euro-Institutes*. Stuttgart: Franz Steiner Verlag, pp. 269–88.

Klein, N. (2003), 'Fortress Continents', *The Guardian*, 16 January.

Kofman, E. and Sales, R. (1992), 'Towards Fortress Europe', *Womens Studies International Forum* 15(1): 29–39.

Kramsch, O., Pijpers, R., Plug, R. and Houtum, H. van (2004), 'Research on the Policy of the European Commission towards the Re-bordering of the European Union', research report prepared for EXLINEA (European Commission's Fifth Framework Programme).

Larsen, J., Axhausen J. and Urry J. (2006), 'Mobilities', in *Mobilities, Networks, Geographies*, Aldershot: Ashgate, pp. 47–62.

Liikanen I. (2008), 'New Neighbourhood and Cross-Border Region Building: Identity Politics of CBC on the Finnish–Russian Border', *Journal of Borderlands Studies* 23: 19–39.

Linde-Laursen, A. (2010), *Bordering. Identity Processes between the National and Personal*, Aldershot: Ashgate.

Löfgren, O. (2008), 'Regionauts: The Transformation of Cross-Border regions in Scandinavia', *European Urban and Regional Studies* 15(3): 195–209.

Maier, C.S. (2006), *Among Empires: American Ascendancy and its Predecessors*, Boston, MA: Harvard University Press.

Meinhof, U.H. (ed.) (2002), *Living (with) Borders: Identity Discourses on East–West borders in Europe*, Aldershot: Ashgate.

Meinhof, U.H. and Galasiński, D. (2002), 'Reconfiguring East–West identities: Cross-generational Discourses in German and Polish Border Communities', *Journal of Ethnic and Migration Studies* 28(1): 63–82.

Mitchell, J.P. (2006), 'Performance', in C. Tilley, W. Keane, S. Küchler, M. Rowlands and P. Spyer (eds), *Handbook of Material Culture*, London, Thousand Oaks, New Delhi: SAGE Publications, pp. 384-401.

Mol, A. (2002), *The Body Multiple: Ontology in Medical Practice*, Durham, NC: Duke University Press.

Mrinska, O. (2006), 'The Impact of EU Enlargement on the External and Internal Borders of the New Neighbours: The Case of Ukraine', in J.W. Scott (ed.), *EU Enlargement, Region Building and Shifting Borders of Inclusion and Exclusion*, Aldershot: Ashgate, pp. 81–93

O'Dowd, L. (2010), 'From a 'Borderless World' to a 'World of Borders': 'Bringing History Back In', *Environment and Planning D: Society and Space* 28: 1031–50.

Orbie, J. (2008) (eds), *Europe's Global Role: External Policies of the European Union*, Aldershot: Ashgate.

Pavlakovich-Kochi, V., Morehouse, B.J. and Wastl-Walter, D. (eds) (2004), *Challenged Borderlands: Transcending Political and Cultural Boundaries*, Aldershot: Ashgate.

Perkmann M. (2007), 'Policy Entrepreneurship and Multilevel Governance: A Comparative Study of European Cross-border Regions', *Environment and Planning C-Government and Policy* 25: 861–79. DOI: Doi 10.1068/C60m.

Rumford, C. (2008), 'Introduction: Citizens and Borderwork in Europe', *Space and Polity* 12(1): 1–12

Rumford, C. (2006), 'Theorizing Borders', *European Journal of Social Theory* 9(2): 155–69

Sandberg, M. (2009a), 'Grænsens nærvær og fravær: Europæiseringsprocesser i en tvillingeby på den polsk-tyske grænse', Ph.D. thesis, University of Copenhagen, Saxo-Institute.

Sandberg, M. (2009b): 'Performing the Border. Cartographic Enactments of the German–Polish Border among German and Polish High-School Pupils', *Anthropological Journal of European Cultures* 18: 107–28.

Schatzki, T.R, Knorr-Cetina, K. and Savigny, E. v. (2001), *The Practice Turn in Contemporary Theory*, New York: Routledge.

Sheller, M. and Urry, J. (2006), 'The New Mobilities Paradigm', *Environment and Planning A*, pp. 206–26.

Schmitt-Egner P. (2001), 'Transnationaler Regionalismus' als Gegenstand der Politikwissenschaft', in J. Bellers and C. Rosenthal (eds), *Die gesellschaftliche Basis von Aussenpolitik. Internationale Wirtschaft, Umwelt, Ideologien, Regional- und Entwicklungspolitik*, Münster: Lit-Verlag, pp. 386–441.

Schultz, H., Stokłosa, K. and Jajeśniak-Quast, D.J. (2002), *Twin Towns on the Border as Laboratories of European Integration*, Frankfurt an der Oder: Frankfurter Institut für Transformationsstudien.

Scott, A.J. (1996), 'Regional Motors of the Global Economy', *Futures* 28(5): 391–411

Scott J.W. (2000), 'Euroregions, Governance, and Transborder Cooperation Within the EU', in M. v. d. Velde and H. v. Houtum (eds), *Borders, Regions and People*, London: Pion, pp. 104–15.

Scott, J.W. and Matzeit, S. (2006) (eds), *EXLINEA Lines of Exclusion as Arenas of Co-operation: Reconfiguring the External Boundaries of Europe – Policies, Practices, Perceptions*, (contract HPSE-CT-2002-00141), http://www.exlinea.comparative-research.net/fileadmin/user_upload/reports_and_publications/Executive_Summary.pdf

Star, S. and Griesemer, J. (1989), 'Institutional Ecology, 'Translations' and Boundary Objects: Amateurs and Professionals in Berkeley's Museum of Vertebrate Zoology, 1907–39', *Social Studies of Science* 19(3): 387–420.

Strüver A. (2004), '"We are only allowed to react, and not act": Eurocrats' Strategies and Borderlanders' Tactics in a Dutch–German Cross-border Region', in B.H. Olivier Kramsch (ed.), *Cross-Border Governance in European Border Regions*, London and New York: Routledge, pp. 25–40.

Strüver A. (2005), 'Bor(der)ing Stories. Spaces of Absence along the Dutch–German Border' in H. van Houtum, Kramsch, O. and Zierhofer, W. (eds), *B/ordering Space*. Ashgate, Aldershot, pp. 207-221.

Szmagalska-Follis, K. (2009), 'Are the European Union's New Boundaries like the Iron Curtain? 1989, Borders and Freedom of Movement in Poland and Ukraine', *International Journal of Politics, Culture, and Society* 22(3): 385–400.

Urry, John (2008), 'Moving on the Mobility Turn', in W. Canzler, V. Kaufmann and S. Kesselring, (eds), *Tracing Mobilities*: *Towards a Cosmopolitan Perspective in Mobility Research*, Aldershot: Ashgate, pp. 13–23.

Van Houtum, H., Kramsch, O. and Zierhofer, W. (eds) (2005), *B/ordering Space*, Aldershot: Ashgate.

Van Houtum, H. and Pijpers, R. (2007), 'The European Union as a Gated Community: The Two-faced Border and Immigration Regime of the EU', *Antipode* 39(2): 291–309.

Van Munster, R. (2009), *Securitizing Immigration: The Politics of Risk in the EU*, Basingstoke: Palgrave Macmillan.

Walters, W. (2006a), 'Border/Control', *European Journal of Social Theory* 9(2): 187–203

Walters, W. (2006b), 'Rethinking Borders Beyond the State', *Comparative European Politics* 4(2–3): 141–59

Wolczuk, K. (2002), 'The Polish–Ukrainian Border: On the Receiving End of EU Enlargement', *Perspectives on European Politics and Society* 3(2): 245–70.

PART I
Europe Re-scaling and its New Security Regimes

Chapter 1
Re-scaling Europe

Michael Keating

Europe of the States

There is a familiar map of Europe, containing some fifty states demarcated by national borders. These borders are, perhaps for the first time in history, largely uncontested, especially in the western side of the continent. Yet there seems little logic or reason to the distribution of states, some of which are large and some very small. Some have seemingly natural frontiers, including mountains or seas while others seem arbitrary. Some have their own languages while others share languages and some are home to more than one language. Some are based on solid national identities while others contain more than one nationality. Yet the social sciences have generally taken these as the basic units for analysis, whether in international or domestic affairs. They have used the term 'nation-state' in an unreflective way to encompass two quite different ideas. One is the notion that the nation corresponds with the state, so allowing us to distinguish nation-states from multinational ones. The other, common in International Relations, refers merely to the sovereign state without any reference to nationality at all; in this case it might be better to talk about the sovereign state or, if states are sovereign by definition, just to the state. In some European cultures, the identity of the nation with the state is so taken-for-granted that the two words can be used synonymously.[1] When the state was largely unchallenged as the normal form of political unit, these contradictions could pass largely unnoticed, except among the inhabitants of stateless nations. Now that the state's hegemony is challenged from above, in the form of transnational integration, from below by new territorial movements, and laterally by the advance of the market, it is more apparent that the nation of the nation-state is rather problematic and that rather than being a single 'thing' it is represents a complex process of territorial bounding that varies greatly from one place and one era to another.

Social science has often tended to teleological accounts of the emergence and consolidation of nation states. Nationali*sts*, of course, look to the awakening of the nation as a natural and inevitable process but national historians, even without an explicitly nationalist bias, have treated the emergence of the nation and state

1 My book *Nations against the State* (Keating, 1996) was translated into Spanish as *Naciones contra el Estado*, but the French translator insisted that Nations contre l'Etat would be an oxymoron, so it was rendered as *Les défis du nationalisme moderne*.

as twin processes to be documented rather than problematized; they have written of the 'unification' of Italy and Germany, for example, as though the nation was already there and merely needed the state to conform to it. Sociology and political science have focused on functional integration and diffusion as processes binding together territories into integrated units, encompassing the state, the economy, society and culture. This perspective is associated with modernization theory, which sees function as coming to prevail over territory as a natural and, often, normatively desirable process. So modernization means territorial integration and functional differentiation. Durkheim (1964) argued that as society become more modern, so territory diminished in importance. Karl Deutsch (1972) showed the nation state expanding to incorporate and integrate territory up to the boundaries of the adjacent project; where distinctiveness remained, the territory concerned would break away to form its own state, so ensuring national homogeneity. These ideas in turn formed the basis for the functionalist and neo-functionalist theories of European integration, applying the same perspective at a higher spatial scale.

From the 1990s, the modernist perspective took a further step in theories of globalization, again driven by functionalist accounts of political order. There are many versions of globalization theory, but the core idea is based on economic integration, cultural diffusion and cosmopolitan political identities, which transcend national borders and reduce the power of nation states. Globalization theorists wrote about the end of territory (Badie, 1995), the borderless world and the end of the nation state (Ohmae, 1995). It is no coincidence that these prognostications of the end of territory coincided the end of the Cold War and with ideas about the end of history as both space and time were swallowed into a single social, economic and political model, defined by US liberal capitalism.

The Persistence of Territory

An alternative perspective sees states and borders and historically contingent and, in many cases, very recent. Osiander (1994) has criticized the Westphalian myth, the idea that the modern sovereign bounded state was created at the treaties of Münster and Osnabruck in 1648. A glance at a map confirms that the only European state to have the same borders in 2011 as in 1648 is Portugal; few borders are older than the twentieth century. Westphalia rather serves as a historical myth, elaborated in the late nineteenth century to legitimize new and hard state borders, and rehearsed in the late twentieth century to define a post-Westphalian order.

The European nation state certainly aspired to create unified territories and defined borders, starting in the early modern period and culminating in the late nineteenth century. The fall of the central empires at the end of World War I confirmed the nation state model as the European norm. Yet both state and nation-building are multidimensional processes, more successful in some respects and some places than in others. States are based on clearly-delineated territories. Within these state-builders seek to construct a national society, internally integrated and externally demarcated; a culture based often on language and always on shared reference points; a national economy; and a system of political domination and representation. Yet these were achieved to different degrees and in different ways, which would leave distinct historical legacies to the states so produced. Some states were created through the integration of diverse territories, often around a central point. In these cases, states were more or less successful in creating nations; for example France was more successful than Spain. Others were created by the fragmentation of empires along national and ethnic lines. Given the contested and fluid nature of ethnic identities and their lack of necessary corresponded to territory, the boundaries of such states have often been in contention.

By the middle of the twentieth century, the nation-state did, for all its difficulties, provide the framework for political and economic life across Western Europe. Internal economic barriers had been removed. National identities had been forged and citizens socialized into them. Political parties usually operated across the state, aggregating interests and taking them to the centre for resolution. Both employers and labour movements were organized on a national scale, bound within the frontiers of the state and so doomed to talk to one another and reach social compromises with each other and the state. This was one basis for national welfare bargains. Another was the shared national identity underpinning redistribution. Integration, however, was never complete, nor was it a one-way or one-and-for-all process.

Stein Rokkan, in his various works (Flora et al., 1999) showed how the process of state consolidation was often uneven across its various dimensions. Economic, cultural and political integration did not always coincide, leaving significant territorial cleavages within and across European states. Others have shown how modernization itself creates new territorial cleavages, through uneven development and the mapping of social and economic struggles onto space. The creation of states, and particularly their bounding in the nineteenth century by hard borders, made territories that were central in older trading orders peripheral within national economies. So the unification of Italy and Germany created serious economic–territorial tensions, given the different interests of their constituent parts in free trade or protection. Spain was industrialized not at the centre but in the peripheral regions closest to the European mainland, an important factor in the rise of Catalan and Basque nationalism.

The European state thus has had to practise more or less elaborate strategies of territorial management in order to maintain its territorial integrity and cohesion (Keating, 1988). These have included socialization of populations into the national ideology, with France showing the way. There are institutional mechanisms, running the gamut from centralization to federalism. Policy responses have included differentiated policies, tariff policies in the nineteenth century, and regional policy after World War II. Regional policy, diverting economic growth within the boundaries of the state could be presented as a non-zero-sum game. The lagging regions could gain from increased investment; the booming regions would see inflationary pressures and congestion reduced; and the national economy would gain from bringing otherwise idle resources into use. Political parties have often made differentiated appeals across the state, so that the apparent homogenization of voting behaviour can give a misleading impression of integration. Italy is a case in point, where the collapse of the old *partitocrazia* has laid bare regional differences and allowed territory to enter into the game of party competition.

The Reinvention of Territory

The arguments in the 1990s about the end of territory often rested upon modernist assumptions about function overcoming space as a principle of social, economic and political organization. They also tended to see the nation state as the defining, if not the only, form of territorial polity. If the state was becoming functionally redundant, it was argued, it should therefore wither away. There was an argument about European integration and whether it represented a deeper, if geographically confined, process of globalization and boundary-suppression or a process of boundary-building in imitation of the state model, but wide agreement that Europe-building would be at the expense of states. This was always to underestimate the political importance of states and the investment which many economic, political and social power-holders had made in them. Milward (1992) notably argued that the purpose of the European project was to rescue the nation state in new global conditions.

The important point for our purposes is that globalization, European integration, and the wider processes of economic change did not suppress territory as such but have in important ways disassociated it from the nation state. We are seeing not so much the end of territory as its re-invention and reshaping. The process can be traced in the domains of economics, culture, politics and institutions.

It is true that the process of economic change can only be understood within a global perspective, given the importance of international trade and corporate ownership. Within Europe, the single market defines a new area of intense transactions and common rules, based on free movement of goods, services, capital and people. Yet there is now a vast literature on the importance of local and regional territories in understanding economic change, the division of labour and the distribution of prosperity. In spite of instant communications, those very sectors that depend on them, such as financial services or computer software production, are intensely concentrated spatially. Without going into the extensive literature on the new economic geography, we can say that the spatial concentration of activity is no longer linked to traditional factors of production like natural resources or proximity to watercourses, but to factors inherent in local societies (Scott, 1998; Crouch et al., 2001). These include tacit knowledge, face-to-face communication and adaptability to change. Some of the literature explains the localization of success as a result of lowered transaction costs. A more culturalist literature attributes it to the inherent virtues of the people, expressed as social capital (Putnam, 1993). My own argument is that it is linked more to institutions and the capacity of local elites to mobilize around shared themes (Keating et al., 2003). In any case, there is widespread agreement that the qualities of territories play a role in explaining growth and that these no longer correspond to state borders.

Globalization and European integration have also reduced the ability of states to manage their spatial economies. Grants, tax allowances and cross-subsidies, key tools used in the past to steer investments to regions, are either illegal or tightly regulated by European law and agreements in the World Trade Organization. Firms refused permission to invest in booming regions can relocate outside the state, or beyond Europe altogether. States themselves, faced with global economic competition, increasingly stress their most competitive locations and sectors, finding regional support programmes too costly. Thinking about spatial development has in any case shifted from an emphasis on the old instruments such as subsidies and hard infrastructure, towards 'softer' factors like human capital, research and development, innovation and the promotion of entrepreneurship. These instruments are often in the hands of local and regional governments. The old idea of regional policy being used to integrate regions within national economies has given way to a competitive regionalism, in which regions compete with each other in European and global space. The extent to which such competition is an objective effect of changing conditions of production, or a political construction, is contested but it has become a major theme in territorial politics. At the same time, wealthy regions see less reason to redistribute resources to their poorer co-nationals. In the past this was part of a territorial bargain in which resources transferred would come back to the wealthy region in the form of purchases for their goods; and redistribution was in any case the price for keeping the state together. Now that wealthy regions see themselves as competing in European and global markets, spatial transfers are seen as more of a burden. So, demands have been rising in the better-off parts of Germany, Italy, Spain and Belgium, to limit equalization payments to poorer regions.

There is also evidence of rescaling of identities and culture. We do not have good, systematic, cross-national survey research on identities, but there is clear evidence that in multinational states like Spain, the United Kingdom and Belgium, identities are divided between attachment to the state and to its component parts, with many people able to sustain dual identities. Parties representing territorial interests, whether nationalist, regionalist or just regionally based, have become an important element in the politics of Spain, the United Kingdom, Italy, Germany and Belgium (where there are no longer any state-wide parties). Minority languages are showing the same tendency as we have noted in relation to the economy, in that globalization and technological advance are encouraging territorial concentration. This is because states have lost the moral legitimacy or the means to suppress such languages, while technology has lowered the costs of production and dissemination. Yet, while it may be easy to imagine global, de-territorialized communities of language communicating via the new media, language communities need proximity for day-to-day informal contact and institutions in education, administration and elsewhere to sustain the life and prestige of the language. So we find minority languages thriving where they have a secure territorial and institutional base. So Basque is doing better within the Basque Autonomous Community of Spain than in the wider Basque area. French in Canada is consolidating in Quebec but under threat outside it. The territorial boundaries of the language communities of Belgium are rigidly fixed.

There is less evidence that social solidarity is rescaling. Welfare state institutions tend to be dominated, if not monopolized, by the states and social citizenship is still identified largely with the nation-state. State-wide welfare bargains, however, have come under strain in recent years, even before the current crisis. The opening of state borders has allowed owners of capital to threaten to opt out of national bargains, while we have noted the tendency of wealthier regions to complain about the demands of territorial solidarity. There is sporadic evidence that new welfare compromises might be emerging at new spatial levels, where there are greater or lesser reserves of social solidarity, and that territory may be, if not replacing, then complementing the nation state as the locus of social solidarity (Keating, 2009).

There is rescaling in politics, as territory becomes an important element in political competition. Territorial parties have entered the arena, both at regional and state level, forcing the state-wide parties to adapt their appeal. This has put on the agenda both the territorial distribution of resources and the question of territorial autonomy.

Conceptualizing Space

We are seeing the emergence or re-emergence of territory at new levels in a world in which the nation-state no longer has a monopoly on its definition. It would be an error, however, to argue that a world, or a Europe, of states is being replaced by one of regions. This is not a world of tightly-bounded territories within which all social, economic, cultural and political process are contained. Modern geographical conceptions of territory have been influenced by the constructivist turn in social science, which emphasizes the way in which the social world is known through organizing concepts. This is not an ontological claim, a denial that the world exists, but an epistemological one, about how we can know it. While working through conceptual schemas, we must not reify them or insist that there is only one way of apprehending social reality. There has followed a series of debates over how one should conceptualize territory. At one end are those who seek to draw lines containing cohesive territories within them. At the other are exponents of the relational or topological approach, who emphasize specific territorially-located activities and the global links among them. There are exponents of the new regionalism who see regions emerging as integrated spaces; and supporters of the global-city perspective, who see a global hierarchy of activities which draws in selected sectors and classes in particular locations. There is a growing realization that none of these provides a master-narrative; they are rather different lenses highlighting different aspects of a complex reality (McLeod and Jones, 2007). This does not make territory a special kind of concept in social science, since we know that other ontological devices like class, ethnicity and gender are subject to the same problems of definition and operationalization.

Adopting a territorial perspective does not mean abandoning other forms of analysis but rather taking a particular angle on the world. It does entail abandoning the older wisdom that the only relevant form of territoriality is that defined by the nation state. We can thus see new spaces emerging in the economy, society, culture and politics, which do not necessarily correspond to the state. They can be supranational (above the state); sub-state; or transnational (crossing the territory of more than one state). The production of such territories cannot be explained simply as the result of functional imperatives, but must be seen as the work also of human agents. Different social interests seek to migrate to new levels where they might expect better treatment, engaging in 'partial exit' from national systems (Bartolini, 2005). So, large firms may seek to evade national regulation by operating at a global level. Trades unions seek to defend the nation-state because they have a legitimate place in the social compromises struck within it and can use it to restrict the mobility of capital. Yet they also seek an entrée at the European level and have used the local and regional level to broaden the agenda of competitive development politics to include social considerations. New social movements, excluded from state-bound systems of negotiation and social compromise, may be attracted to new and emerging territorial levels. Some groups may be fortunate enough to operate at several levels and so can venue-shop for the level where they will be treated best. This means that the constitution of new territories is a matter of social and political contention.

Institutionalizing Space

The various processes that we have traced have created a variety of new functional, social, political and cultural spaces, which may or may not coincide. The extent to which they do is not itself the result of functional restructuring but of political agency, the strategies of actors to create new institutional forms and endow them with capacity. A common metaphor to describe these emerging forms of regulation and collective action is 'governance' or, at the spatial level 'multilevel governance'. This is a term with so many different meanings that its analytical value must be seriously doubted but the most common seems to refer to a system of policy-making characterized by horizontal coordination, networks and negotiation, rather than hierarchy and authority. Sometimes we are told that the world in general is moving from government to governance, a claim that is difficult to test empirically and which seems to rely on a highly stylized view of old forms of government. Very commonly the term is used for policy making at the sub-state and at the European level, presumably because the weight of the old nation-state is less preponderant. The term has also been instrumentalized by states, international organizations and the European Union to suggest a softer, gentler way of managing the public domain, free both of the brutality of markets and the domination of government.

All of these meanings tend to minimalize social conflict and politics, and to point towards consensual forms of collective action aimed at common goods.

This is profoundly misleading. The new territorial spaces, whether at the supranational, sub-state or transnational levels, are sites for politics and clashing interests. Not only is there politics within the new spaces, but the shape of the new spaces is also a matter of political debate. This contestation has given rise to rival projects for the regulation and politicization of new spaces. The European Union has sought to valorize transnational spaces through cross-border and inter-regional cooperation programmes, providing resources and institutional support but the variety of interests at stake often makes it difficult to establish common goals and priorities. State governments have sought to regulate new spaces through establishing regional institutions aimed at managing problems that cannot be handled at the central level. These include economic development, the management of health, active labour market policies and training, planning and infrastructure. Social movements and opposition parties have sought to politicize the new spaces through the creation of elected governments and to broaden the regional agenda. Business interests have often supported the idea of a regional level, for efficiency and functional reasons, but prefer appointed agencies or corporatist institutions in which they have a privileged role, without being challenged by social interests or environmentalists and with a rather narrowly-defined development agenda. Trades unions, traditionally rather centralist, have increasingly invested in regional institutions as a new level for influence, compensating for the loss of power at the workplace, and have sought to broaden the political agenda.

The result is a variety of institutions at varying spatial scales from the local to the supranational but expressing two tendencies. One is towards a more general-purpose form of regional government at the meso (intermediate) level in all the large states – Germany, Italy, France, Spain, Poland and the non-English parts of the United Kingdom – as well as Belgium. These have varying degrees of power and autonomy from the French regions, which are functionally circumscribed and matched by a regional tier of national government, to the powerful general-purpose Länder in Germany. There has been a gradual progression from forms of regional administration, agencies and corporatist forms of cooperation around development, towards general-purpose regional government as states seek to establish more effective institutions and to offload difficult spending choices, while social movements seek to broaden the agenda through direct election. England is the only place in Europe where this process has not only been stalled but reversed and not only on one but on two occasions, in the 1960s–70s and in the 1990s–2000s. Hungary has stepped back from elected regional government, although committed in principle.

The other tendency is towards consolidation at the metropolitan level, around the big cities. There is a tension here between the need for action at the larger scale, encompassing the wider functional area, and local political interests entrenched at a lower level. There is an intellectual division between exponents of municipal consolidation to create larger units, and public choice people, who favour fragmentation. There are also conflicts between wealthier districts, whose inhabitants wish to keep their locally-generated tax revenues, and poorer ones more favourable to redistribution. The outcome is typically a compromise, in which metropolitan governments are established but are indirectly elected from the lower tier and with restricted powers and resources.

Europe of the Regions

Europe has provided a new context for territorial restructuring. This is not, as some observers once thought, because the European Commission has a plan to subvert nation-states but because of the functional and institutional dynamics of the integration project. The internal market, as noted above, tends to accentuate inter-regional competition. It may also increase territorial disparities; at one time this was captured in the image of the 'golden triangle', the old industrial core, later replaced by the 'blue banana'.[2] More recently, the predominant image is of the mosaic, with islands of prosperity and poverty across European space. These disparities provoked some opposition to the European project in peripheral and lagging regions, leading to the EU instituting its own spatial policies, now subsumed into the Structural Funds or cohesion policy and the inclusion of an objective of territorial cohesion in the Lisbon Treaty. The Structural Funds have not in themselves encouraged the institutionalization of regions, as they define problems and dispense funds at a variety of territorial levels, but they do represent a commitment to a degree of spatial equity and have been the vehicle for transmitting the new ideas about regional development.

A second effect of European integration has been on the workings of federal and devolved systems of government. Since it is the member state that is represented in the Council of Ministers, national governments have been able to re-enter devolved policy fields where they are Europeanized. The response to this has been a series of initiatives to give regions more institutional presence in Europe. Under certain conditions, regional ministers can represent their state in the Council of Ministers, a provision used by Germany, Austria, Belgium, the United Kingdom, Spain and Italy. Sub-state governments have also been given recognition in the advisory Committee of the Regions. This has not, as was once sometimes thought it might, developed into an additional legislative chamber, or indeed a significant EU institution, mainly because of its heterogeneous membership, including both municipal governments

2 The golden triangle metaphor was popular in the 1960s and described an area from the industrial Ruhr, through south-east England to Paris. The blue banana, elaborated by French economists in the 1980s, went from south-east England, through the Low Countries and continued in an arc to northern Italy. It bears a strong resemblance to the trading routes of the early modern period, when it was marked by the presence of trading city-states and regions.

and federated regions. Since there is no commonly-agreed 'regional' level across Europe, it is not possible to institutionalize it in a consistent way.

A third effect of Europe is to provide minority nationalist movements within states with a new framework for their autonomy demands. For some, like the Scottish National Party, Europe provides an external support system for statehood, dealing with externalities and difficult problems like security of markets, borders, the currency and, potentially, security. Others have used the example of Europe to move beyond traditional notions of state sovereignty altogether, arguing for a confederal Europe in which power can be divided at multiple levels.

Fourth, Europe has in some respects consolidated borders, since borders between EU states are largely unchallenged. At the same time, it has emptied them of part of their meaning. They are no longer economic barriers nor, as long as Schengen holds, obstacles to the movement of people. They are, however, political frontiers and demarcate citizenship and belonging. The border itself then becomes transformed, serving to efface some meanings of territory while affirming others. Within border regions, there is also a competition for the definition and domination of space, often pitching the European Commission, with its commitment to functional integration, against national governments and regional and local governments, operating at varying spatial scales and answering to different constituencies.

Regional Futures

The often exaggerated ideas about Europe of the Regions that were about in the 1990s can now be discounted. States have proved resilient and play a major role in defining and institutionalizing space while the protagonism of the European Commission, in this as in other fields, is not what it was. Regions and localities are too heterogeneous in their construction to fit into a simple model or a three-level scheme of federalism. Yet territorial rescaling continues across the continent. Spain, Belgium and the United Kingdom face powerful nationalist movements seeking a new territorial order, which seems more likely to end up in a kind of confederal arrangement than lead to classic statehood. Territory is a powerful influence on economic change and territorial disparities remain an important issue within the single market. Faced with real or imagined threats of terrorism or of migration, states have recently sought to strengthen their borders. The United Kingdom has established a massive and intrusive regime of border control,[3] while the Schengen agreement has been curtailed. Yet the re-establishment of controls on movement is economically costly and governments know it.

3 A Kafkaesque incident recently brought this home to me when I was invited to examine a doctoral thesis at the University of Glasgow and I, a UK citizen living in the UK and working at a UK university, had to bring a passport to satisfy the UK Border Agency that I was eligible to work in the United Kingdom.

In some parts of Europe, new territorial systems have emerged, combining institutions, identities, economic development and social solidarity, in places like Flanders, Catalonia or Scotland. Elsewhere, there are region- or nation-building projects but the territory or its significance is challenged, as in the Basque Country or northern Italy (Padania). City-regions are emerging as important economic, social and political spaces in many countries. Elsewhere, there is little or no region-building and territories remain subordinate to state policy and strategy. Europe is rescaling but is doing so in very different ways in different places.

References

Badie, B. (1995), *La fin des territoires. Essai sur le désordre international et sur l'utilité sociale du respect*, Paris: Fayard.

Bartolini, S. (2005), *Restructuring Europe*, Oxford: Oxford University Press.

Crouch, C., le Galès, P., Trigilia, C. and Voelzkow, H. (2001), *Local Production Systems in Europe. Rise or Demise?*, Oxford: Oxford University Press.

Deutsch, K.W. (1972) *Nationalism and Social Communication*, Cambridge, MA: MIT Press.

Durkheim, E. (1964), *The Division of Labour in Society*, New York: Free Press.

Flora, P. (ed.) (1999), *State Formation, Nation-Building and Mass Politics in Europe. The Theory of Stein Rokkan*, Oxford: Oxford University Press.

Keating, M. (1988), *State and Regional Nationalism*, Brighton: Harvester-Wheatsheaf.

Keating, M. (2001a), *Nations against the State. The New Politics of Nationalism in Quebec, Catalonia and Scotland*, 2nd edition, London: Palgrave.

Keating, M. (2009), 'Social Citizenship, Solidarity and Welfare in Regionalized and Plurinational States', *Citizenship Studies*, 13(5): 501–13.

Keating, M., Loughlin, J. and Deschouwer, K. (2003), *Culture, Institutions and Economic Development. A Study of Eight European Regions*, Cheltenham: Edward Elgar.

MacLeod, G. and Jones, M. (2007), 'Territorial, Scalar, Networked, Connected: In What Sense a "Regional World"?', *Regional Studies* 41(9): 1177–91.

Milward, A. (1992), *The European Rescue of the Nation State*, London: Routledge.

Ohmae, K. (1995), *The End of the Nation State: The Rise of Regional Economies*, New York: The Free Press.

Osiander, A. (1994), *The States System of Europe, 1640–1990: Peacemaking and the Conditions of International Stability*, Oxford: Clarendon.

Putnam, R. (1993), *Making Democracy Work: Civic Traditions in Modern Italy*, Princeton, NJ: Princeton University Press.

Scott, A. (1998), *Regions and the World Economy*, Oxford: Oxford University Press.

Chapter 2
Forcing Flows of Migrants: European Externalization and Border-induced Displacement

Martin Lemberg-Pedersen

Introduction

The European borders and the practices and technologies that constitute them are contested. This chapter examines the development of EU border control with a particular focus on its external dimension as it is manifested towards African countries and migrants. Since the categorization of cross-border mobility is a disputed domain, I invoke the concepts of refugees, asylum seekers and migrants during the chapter. At the risk of conceptual ambiguity, this is nevertheless done in order to show how discursive identifications can activate various regimes of enforcement towards human mobility.

At the outset it is argued that international negotiations, external governance and extra-territorialization are all aspects of externalization and they are illustrated through the examples of the European Neighbourhood Policy [ENP], the Tampere and Seville Presidency Conclusions and the Frontex Agency's HERA-operations. The notion of borderscapes is then suggested as a useful analytical tool for understanding the multiple abstractions of knowledge, practices and technologies at work in EU border control highlighting the fact that the EU borders change over time according to political and administrative processes. Since border control re-territorializes geographic spaces according to the mobility of the people through them it has substantial geopolitical aspects (Deleuze and Guattari, 1987; Walters, 2004). Such a perspective on European borderscapes, however, cannot stand alone, but must be complemented with the more governance and process-oriented analytical perspective of biopolitics (Balzacq, 2009; Rajaram and Grundy-Warr, 2007; Vaughan-Williams 2009). It is, however, argued that a critical focusing on sovereign power and black letter law bypasses the bureaucratic and political actors involved in the depoliticization and securitization of European asylum policies (van Munster 2009; Bigo, 1996; Chimni, 1998).

It is claimed that the geo- and biopolitical instrumentalizations reflect a power geometry (Massey, 1993) realized through an infrastructure underpinned by continuous corporeal, financial, material and virtual flows designed to yield European hypermobility and non-European submobility (Hyndman, 2000) and to contain displaced people before they reach European territory. It is shown that these complex flows take different forms, such as European Commission [EC] donations to the United Nations High Commissioner for Refugees [UNHCR] and programmes supporting third countries migration control, the deployment of Immigration Liaison Officers [ILOs] and Frontex operations and bilateral military agreements. Moreover, it is pointed out that the funds underpinning the European borderscapes are not only flowing from European to third countries, but also increasingly to Private Security Companies [PSCs] (Gammeltoft-Hansen, 2011; Lemberg-Pedersen, 2012).

Finally, the chapter suggests that works on forced migration have tended to view border control as a reaction to the movement of already-displaced people, but that externalization in itself should be seen as a cause of displacement and forced migration. This allows us to appraise the overall consequences of the external EU borderscapes' decentralized functionality instead of particular border control events. Conceptualizing the EU borderscapes' external dimension as the sequencing of forced flows makes it possible to identify its production of border-induced displacement and highlights the interrelated functionality of the transnational European borderscape thereby providing a lens through which the extent of its humanitarian consequences can be grasped.

The External Dimension of the European Borderscapes

During the 2000s the concept of externalization has been invoked as a description of how European states have exported aspects of migration control to third countries (Boswell, 2003; Betts and Milner, 2007). To motivate the drive for external measures the EU's 'Strategy for the External Dimension of JHA: Global Freedom, Security and Justice' (Council of the European Union, 2005) notes that in order to respond to the security threats of terrorism, organized crime, corruption and unmanaged migration flows, the 'development of the area of freedom, security and justice can only be successful if it is underpinned by a partnership with third countries on these issues which includes strengthening the rule of law, and promoting the respect for human rights and international obligations.' Externalization, it seems, is realized in different ways.

The concept of borderscapes captures how the European border control is not a static, geographical phenomenon, but dynamic, consisting of political power, technological practices and knowledge-production. Borderscapes can be defined as multidimensional abstractions of knowledge and technologies (Rajaram and Grundy-Warr, 2007). It highlights the fact that borders are fluctuating landscapes always in the process of being constructed (Rancière, 2004). Viewing EU externalization as a complex process of borderscaping therefore allows us to open up the border as a sociogeographic space, underpinned by varying technologies, infrastructures and actors.

Thus, the EU or its Member States negotiate specific agreements related to migration control with third countries. Such negotiations can result in the union attempting to regulate, manage or control the asylum and migration policies of third countries through external governance. This, in turn, may then yield instances of extra-territorialization, where material conditions for controlling migratory movements are put in place beyond European territory (Balzacq 2009, pp. 2–3). As such external governance can result from negotiations between the EU and African countries with the consequence that structures of migration control emerge where they did not exist before.

Examples of these forms of externalization abound, but I will here focus on the ENP, Seville's migration-development nexus and the operations of the Frontex Agency. Firstly, the ENP is manifested through negotiations of Action Plans and Association Agreements between the EU and its neighbours. African countries granted the ENP-status include Tunisia, Egypt, Morocco and Algeria while Libya has special arrangements. Part of the ENP-agenda is to bolster these countries' capacities for migration control. For instance, the EU-Morocco ENP agreement meant that the EU committed €40 million to the construction of an advanced radar system to detect immigrants and drug trafficking along Moroccan northern coasts. While this radar system is operated by Morocco, it illustrates how EU-Morocco negotiations resulted in the creation of an extra-territorialized control structure. Secondly, the Migration–development nexus formulated by the Seville Council in 2002 can be seen as a case of external EU governance. This nexus makes the acceptance of the EU's priorities regarding asylum policies a necessary condition for non-EU states if they wish to continue cooperation with the union and receive its development support (Council of the European Union, 2002). Its demand of acceptance therefore effectively regulates the available asylum policies of non-EU states according to EU interests.

Thirdly, Frontex operations also exemplify the extra-territorialization of control practices. Thus, between 2006 and 2007 and following negotiations with Senegal, Mauritania and Cap Verde, Frontex was allowed to conduct the HERA-operations on these countries' territories. The HERA-operations targeted boat migrants seeking to reach the Canary Islands, where they would be able to apply for asylum in Spain. To counter this movement Frontex deployed Spanish helicopters, naval vessels from Italy, Portugal and Spain and aircrafts from Finland and Italy in the territorial waters of Senegal, Cap Verde and Mauritania. Moreover, two surveillance aircrafts from Italy and Finland were flown deep into African territory to monitor migration routes through the deserts. HERA combined interceptions in the territorial waters of these countries with flights over their vast desert areas. According to Frontex this was done as a way of 'preventing them to risk their lives on the dangerous journey' (Frontex website), but also prevented the migrants from reaching Spanish territory and filing asylum claims. In total, the HERA-operations intercepted and diverted around 6,000 boat migrants.

The negotiations facilitating external governance and extra-territorialization illustrate how EU externalization transcends the union's geographical borders and consists of multifaceted processes, which are integrated into the union's relations with third countries. As a form of geopolitics externalization effectively re-territorializes the spaces of third countries according to European interests (see also Walters, 2006, pp. 678–9). The externalized borders are thus mechanisms of social regulation manifesting differentiated landscapes of power by controlling the mobility of people through territories via such practices as surveillance, interception, detention and deportation. This highlights how the external dimension of the European borderscapes has geopolitical and biopolitical components.

The Circulation, Exclusions and Processes of Externalization

Biopolitics can be defined as the instrumentalization of biological processes according to political interests and biopolitical analyses therefore enquire into the 'microphysics of power' manifesting governmental power. Thus, while a geopolitical perspective asks 'why' power is manifested in certain ways, and what interests guide it, a biopolitical perspective asks 'how' this power is realized (Vaughan-Williams, 2009). This brings to the foreground the many concrete instruments and practices through which the macro-perspective of geo-power is pursued. Accordingly, scholars have attempted to deconstruct the technological regimes behind European border and migration control (Balzacq, 2009; Bigo and Guild, 2005) building on the theorizing of biopolitical governmentality done by Michel Foucault (Foucault, 2008, 2007).

According to Foucault (2008), governmentality is a distinct political economy whereby power is delegated and decentralized in order to facilitate the free circulation of people and goods. Both the EU's Schengen Area, with its free movement of EU citizens and goods between union Member States, and externalization delegating control capacities to third countries, can be seen as functioning according to this rationale. Hence, governmentality is not a rationality of absolutely free movement, but functions through the regulation of individual freedom *vis a vis* the overall utility of circulations to the population (Foucault, 2008, pp. 42–3). The governmental ideal of free circulation of flows therefore requires the pre-emption, through security apparatuses, of other, risky, flows. From its inception in 1985, the Schengen Convention, too, has been premised on a link between the free movement of people and compensatory 'flanking measures' safeguarding the internal European space against threatening and risky flows (van Munster, 2009; Boswell, 2003).

The security apparatuses composing these flanking measures are grounded in political, legal and technological practices, but in 2005 a brutal example of what consequences they may have occurred in Ceuta and Melilla, two Spanish enclaves in Morocco, which constitute some of the EU's southernmost borders. Here, Moroccan and Spanish security forces fired upon migrants attempting to reach Spanish territory from both sides of the border-fence. The result was the deaths of 15 people from shots and falls and the wounding of a further 100

(Goldschmidt, 2006, pp. 1–2). Over the following weeks, and after considerable European pressure, the Moroccan authorities dumped 1000 irregular migrants in desert areas near the Algerian border and without food, water or medicine (Doctors without Borders, 2006). After this the geographical sites of Ceuta and Melilla were thoroughly transformed into militarized zones with high-tech surveillance technologies and three lines of fences whose height were increased to 6 meters. Also, the €40 million Moroccan radar system funded by the EU was launched in this period. Ceuta and Melilla are therefore prime examples of how borderscapes can undergo change. The Ceuta and Mellila incidents also open the door to the 'dark side of bio-politics' (Dean 1999, p. 139; Dean 2002, p. 41), showing that the differentiation of free from forced flows through security apparatuses can have tragic consequences.

A biopolitical framework complementing the notion of governmentality and casting further light on these dynamics of exclusion is that of Giorgio Agamben (Agamben, 1995, 1998, 2005). Unlike Foucault's sub-legal analyses of power, Agamben premises his biopolitics on states' juridico-institutional sovereignty in an attempt to call into question the fundamental categories of the nation-state (Agamben, 1998, pp. 9 and 134). Since border control is a prime example of sovereign power, his work lends itself to analyses of asylum policy. Agamben defines sovereign power through its capacity to declare a state of exception whereby human existences are banned, or excluded, from societal status and protection (Agamben, 1998, p. 181). He views the codification of lawful communities as fundamentally functioning through the exclusion of some existences, an exclusion, he calls the biopolitical production of 'bare life'. As border control consigns asylum seekers to exist in zones of exception on the margins of communities they incarnate this bare life (Agamben, 1995, 1998).

In the state of exception sovereign power is expanded to include those that it exclude, in what is termed inclusive exclusions (Agamben, 1998, pp. 104–11) and externalization may be seen as following this dynamic as it is externalized outwards to include those that it excludes under its mechanisms of control. Asylum seekers can therefore be seen as being placed in a relation of inclusive exclusion. Agamben finds the most paradigmatic example of this governance-through-exception in the technology of the camp, because camps by definition blur the relationship between the norm and the exception. In camps, exceptional biopolitics is realized as a normal condition and, accordingly, it functions as a 'line without extension' separating political existence from bare life (Agamben, 1998, pp. 122, 148).

To be turned into bare life is an inherently vulnerable condition and the deplorable conditions in the migrant detention camps within and beyond European territory seem to lend relevance to this concept. For instance, Doctors without Borders (2009) found 140 women and 70 children shoved into a 200 square meter room with only two toilets in the now closed Pagani camp in Greece. Similarly, Human Rights Watch (2009, p. 74) assess the camp conditions in Libya as ranging from 'negligent to brutal' and according to an anonymous diplomatic source the duration of migrants' arbitrary detention in Libya varies 'from a few weeks to 20 years' (Ibid.). In 2009, Ghedi, a 29-year-old Somali told Human Rights Watch about the conditions in Libyan camps:

> The guards beat us for no reason. They would slap us all the time. Once I was beaten with a butt of a rifle. The guards especially came to beat us at night when they were high from smoking hashish. Some guards kicked with their boots, others hit with their hands or with a weapon. We would just be sleeping before the beatings. We didn't make any demands. We did nothing. I was beaten all over my body. (Human Rights Watch, 2009, p. 76)

The Agambenian biopolitics theorizes the separation of migrant flows through such conditions as relations of inclusive exclusion produced by sovereign decisions on the landscapes of power. However, such a perspective faces difficulties both when it comes to pinpointing the complex processes shaping the enforcement of border control as well as assessing migrant agency.

Conceptualizing migrants as instances of bare life is problematic because it reifies the desire of states to reduce migrants to passive existences. Yet, although migrants may be exposed to exploitation and abuse in camps they are not passive existences to be moulded in the hands of European or African authorities. Migrants employ various strategies to manoeuvre within the border networks, such as the bribing of border guards, working irregularly, being employed by smugglers to sail migrants across seas and migrants have staged numerous demonstrations protesting their detention conditions. Consequently, they are agents with the capacity to act in order to better their lives (Lucht, 2011; Bakewell, 2008). Furthermore, the notion of bare life is unable to distinguish between the different conditions migrants experience in different countries. Conceptually speaking, their existence should therefore instead be seen as placed on a biopolitical continuum of inclusive exclusions where the degree of vulnerability depends on the degree of force applied to their mobility.

Moreover, conceptualizing dynamics of exclusion as a binary relation between law and its suspension makes sovereign power an underdetermined abstractum and does not address the sub-legal processes and the actors involved in its manifestation (Lemke, 2005, p. 8). Yet, borderscapes are comprised by multiple actors, technologies, and political interests. In other words, borderscapes undergo reconfigurations reflecting the shifting political paradigms guiding their rationalities. The development of the EU Dublin Convention exemplifies this.

European asylum policy can be seen as having shifted from the use of refugees as pawns in Cold War politics to their current containment in regions of origin (Chimni, 1998, p. 350; Hyndman, 2000; Bakewell, 2008). The Refugee Convention was formulated within the larger political context of the Cold War where European states were eager to demonstrate the openness of liberal democracies to asylum seekers, which were victims of communist totalitarianism. However, as flows of asylum seekers began to come from the decolonizing regions during the 1960s and 1970s and the bipolar world order approached its end this attitude began to change.

The European states reacted not by breaching the Refugee Convention, but by refusing to receive asylum claims. The result was the creation of 'refugees in orbit', that is, refugees without a country where they could seek asylum (Paludan, 1981, pp. 70–71). In 1990 the EU presented the codification of the Dublin Convention as a solution to this problem via the concept of 'country of first arrival'. It was framed as ensuring that 'applicants for asylum are not referred successively from one Member State to another without any of these states acknowledging themselves to be competent to examine the application for asylum' (Dublin Convention, 1990).

While the notion of country of first arrival did place responsibility for the asylum process on a specific country, it also relieved all subsequent countries of arrival from the obligation to review asylum claims. This division of responsibility was used by the EU's Northern Member States to systematize the use of administrative deportations of asylum seekers, since most asylum seekers could be found to have passed through the territory of Southern Member States like Greece or Italy. Moreover, Dublin also created incentives for states to block refugees' access to asylum before they reached their territories. This was done through such practices as pre-arrival screening, visa requirements and carrier sanctions and the result was the 'engineered regionalism' of refugee flows (Gibney, 2007), that is, the re-territorialization of refugees' regions of origin in order to contain them outside European territory. Rather than solving the problem of refugees in orbit, then, the Dublin Convention ended up systematizing the creation of irregular migrants (van Munster, 2009, p. 32).

Invoking essentializing concepts such as sovereign decisionism and states of exception to account for the emergence of Dublin-deportations bypass the webs of power and resistance involved in formulating borderscapes and end up reaffirming states' claims to be able to draw a line between norm and exception (Bigo, 2007, p. 4, 12). Hence, several scholars have argued the need to factor in the bureaucratic processes through which the European borders were recoded (van Munster, 2009; Huysmans, 2006; Bigo, 2007, 2001, 1996). Van Munster argues that this reconfiguration of the European border politics was facilitated by processes of 'bureaucratic crowd-out' between the first and the second Schengen Conventions (in 1985 and 1990 respectively). These bureaucratic turf wars transferred border control competences from foreign affairs and transportation officials to officials from the justice and home affairs departments with internal security as their professional disposition resulting in the gradual 'colonization' of the union's foreign policy by issues of internal security (Lavenex and Ucarer, 2004)

Alongside this transfer, other structures, formal and informal, such as Trevi, the Ad Hoc Working Group on Immigration (AHWGI) and the Coordinators Group were also influential by framing Dublin as a tool to prevent abuse of asylum systems such as the so-called 'asylum shopping'. While both refugee scholars and NGOs contested this reconstruction of the borders, Chimni (1998, p. 355) has argued that the legal positivist perspective from which this critique was levelled meant that is was ill-equipped for countering the sub-legal processes through which the borderscapes were recoded. Viewing international law as an objectively identifiable system of rules separated from the political sphere, refugee scholarship had, effectively, disarmed itself when it came to resisting the expansion of the containment agenda (ibid.).

Together, then, attention to the geopolitical, biopolitical and bureaucratic processes in the European borderscapes help nuance our understanding of the externalization agenda's development. Moreover, it sheds light over the kind of existences that borderscapes make of migrants. Nonetheless, while such perspectives go some way in untangling the processes underpinning borderscapes, we still need to ask about the processes realizing its instances of inclusive exclusion. More specifically, it is necessary to enquire into the roles played by transnational flows of capital, equipment and people in the European borderscapes and the actors facilitating them.

Transnational Flows I: Funds underpinning the Externalized European Borderscapes

The EU's differentiation of circulatory flows exhibits a certain 'power-geometry' where different groups have different relationships to mobility: 'some are more in charge of it than others; some initiate flows and movement, others don't; some are more on the receiving end of it than others; some are effectively imprisoned by it' (Massey, 1993, p. 61). Following William Walters (2004, p. 678) we can see this differentiation as functioning according to a geostrategy of the networked border. While we should acknowledge that the EU borderscapes do not accord fully with any one geostrategy, but always has aspects of several in its rationality, the networked border nonetheless fits with Schengen's cancellation of fixed points of control within the EU and their replacement by a transnational network of flanking measures. Schengen's networked borderscape is diffuse, decentred and de-territorialized (Massey, 1993, p. 681) and its infrastructure therefore rely on the constant circulation of financial, material and corporeal flows. In the following I will consider four such flows, namely the dispersal of funds to UNHCR operations and EU programmes of building third countries' control capacities, the transfers of personnel and equipment to Frontex-operations and third countries and the outsourcing of border control to PSCs.

Because the European migration control is diffused and deterritorialized, it can also take place in venues not normally considered border management. One of these venues is the UNHCR because it is grounded in a state-based structure making its capacity dependent upon the political and economic priorities of its member states. This is illustrated by the fact that state donations are voluntary and how the biggest donors earmark funds to specific projects in specific regions. In 1994, more than 95 per cent of UNHCR's donations came from fourteen governments of industrialized countries and the EC (Hyndman, 2000, p. 57). Two related implications follow from this: If the UNHCR prioritizes areas conflicting with the interests of donor states it faces threats of decreased funding. Conversely, if the UNHCR pursues projects such as repatriation, resettlement or camps in refugees' regions of origin, which promote European hypermobility and non-European submobility (Hyndman, 2000, p. 37), it is likely to attract substantial funding (Hyndman, 2000, pp. 3–5). Although UNHCR staff on the ground may not side with the actors attempting to engineer regionalism, its structural dynamics therefore opens it up to processes whereby powerful donors utilize the agency for purposes of migration control.

Earmarking of funds to specific regional projects therefore has the consequence of turning the agency's need for funding into a motivation for pursuing the geopolitical interests of donors. Realizing this, the European Commission (EC) has followed the US strategy of earmarking *all* of its donations to specific priorities. Reflecting the growth of the UNHCR budget, the EC has almost doubled its donations from 2001 ($66 million) to 2009 ($127 million) making it the second largest donor in the world. Consequently, the flows of funds coming from the EU yield a powerful influence on the overall UNHCR capacity. In 2004 the EC earmarked its donations to projects in those countries producing the most refugees, such as Sri Lanka, Afghanistan, Iran, Ukraine, Pakistan and Thailand, and in general the entire African continent. These projects typically involved assisted returns or voluntarily repatriation of refugees, consolidating or relocating of refugee camps, capacity-building for protecting returnees, building and strengthening of asylum systems in third countries, and the pre-screening of asylum seekers and migrants (UNHCR Global Report, 2004). It can thus be said that the UNHCR plays an important indirect role in the EU's creation of 'space and distance' to displaced people in Africa illustrating how migration control may also be pursued through humanitarian venues.

Other example of transnational financial flows in the European borderscapes are the recent financial EU programmes such as Odysseus, Argo, Aeneas, Meda, Cards, ENPI, DCI and the Border, Return, Refugee and Integration Funds of the 'Solidarity and Management of Migration Flows Programme'. These programmes focus especially on the issues of asylum, immigration and the cross-border movement in third countries experiencing large influxes of European-bound migration. The emergence of these programmes illustrate the EU's increased political focus on the external dimension of its borderscapes as does the fact that their budgets have all grown drastically. For instance, while Aeneas (Programme for financial assistance to third countries in the area of migration and asylum) distributed €20 million to projects between 2001 and 2003 the programme was allocated €250 million between 2004 and 2006 before it was discontinued and replaced by the DCI which, between 2007 and 2013, is tasked with distributing €387 million.

Examples of Aeneas projects include the 'Across Sahara' I and II, which were funded from 2005 to 2009. They were implemented by the Italian Ministry of Interior and the Department of Public Security in collaboration with Niger and Libya. Their objective was to 'fight' and 'combat' irregular migrants from Sub Saharan Africa attempting to reach the coasts of Southern Italy (Aeneas, 2006, p. 5). The EC covered €2.6 million out of the projects' overall budgets of €3.2

million. Other Aeneas projects like 'Project Seahorse' and 'Seahorse Network' (2005–2008) in Morocco, Mauritania, Senegal and Cap Verde were supported with €4 million out of €5,1 million (Aeneas, 2006, pp. 12, 20).

Although some Aeneas projects were concerned with protecting women's rights or strengthening protective measures for asylum seekers, the main focus of the programme accorded with the containment agenda. Thus, while Aeneas supported 14 projects with €14 million in the subsectors of legal migration and asylum and protection, this support was dwarfed by the programme's support to the subsectors of border management, irregular migration and migration management. Here, 24 projects were supported to the tune of €29 million. Out of these €14.4 million were allocated to the subsector for fighting irregular migration alone (Aeneas, 2008). Moreover, the countries where most Aeneas projects were implemented, namely Morocco (20 projects), Tunisia (10 projects), Libya (9 projects), Senegal (8 projects) and Egypt, Algeria and Mauritania (7 projects), are all located along the three main migratory routes towards Europe; the Western African, the Mediterranean and the Eastern Mediterranean routes. The financial flows of the Aeneas Programme can be seen as complementing the EC donations to the UNHCR and exhibit a similar geographical asymmetry since focusing on transit-regions rather than regions of origin.

Both financial flows illustrate how transnational financial flows are vital for the networked EU borderscapes. Yet, European countries are not only countering migration flows by funding third countries' migration management and border control. They also circulate personnel and military equipment through the deployment of ILO's, Frontex-operations, and the donation or sale of military and surveillance equipment via bilateral deals.

Transnational Flows II: Personnel, Equipment and Private Actors Underpinning the Externalized European Borderscapes

Council Regulation 377/2004 codified the creation of ILO networks in third countries (Official Journal of the European Union, 2004). ILOs exchange information with the authorities of these countries and make risk analyses about flows and routes of irregular migrants, their means of transports and the intermediaries facilitating such flows (Official Journal of the European Union, pp. 2–3). The establishment of ILO networks beyond European territory seems to fuse together the networked border with another geostrategy, namely that of the march placing Europe under siege (Bigo, 2000, p. 68; Walters, 2004, p. 686). While the figure of the march long predates the current technological practices of the networked EU borderscape, its reappearance is facilitated by the organizational possibilities offered by the decentralized network. In the current European geostrategy the ILO function is thus viewed as a first bulwark providing deflection measures against flows of migrants embarking onto Europe. The connected functionality of the different practices in the external European borderscape, like ILOs and the reconstructed borderscapes of Ceuta and Mellila illustrate how different geostrategies 'are not mutually exclusive' but complement each other in 'a shifting ensemble of heterogeneous political rationalities and practices' (Walters, 2004, p. 693).

The joint Frontex-operations at the EU's external borders or on third country territory instantiate another transnational flow as they depend on Member States' temporary loans of naval vessels, helicopters and aircrafts. Thus, the HERA operations, taking place in the territorial waters of Senegal, Mauritania and Cap Verde between 2006 and 2008, involved the loans of helicopters, naval vessels and surveillance aircrafts from Italy, Spain, Portugal and Finland. Around 6.000 boat migrants were intercepted while attempting to exit from the African territorial waters and diverted back without being able to access asylum procedures (Frontex-website).

On a bilateral level, individual Member States also transfer military equipment to third countries. Helmuth Dietrich (2004) explains the case of Germany:

> The German government is also responsible for arming the North African coast. According to the German defence ministry, Tunisia will receive six Albatross speed boats from the German navy. Already two years ago, it was agreed to deliver five speed boats to Egypt. In 2002, Algeria received surveillance systems at a value of 10,5 mio €, Tunisia received communications and radar equipment for around 1 mio €, Morocco received military trucks worth 4,5 mio €. (Dietrich, 2004, p. 7)

Similarly, Italy has donated 6 police patrol boats, patrol vehicles and Unmanned Aerial Vehicles (UAVs, or drones) to Libya for the monitoring of the country's southern deserts for migrants (Ibid.). Netherlands exported three SIGMA frigates from to Morocco in 2008, France supplied Algeria with 9 light helicopters in 2006 and the UK supplied four Super Lynx-300 helicopters to the same country between 2004 and 2007 (SIPRI, 2009, p. 13, 2010). The result of the Frontex-circulation and bilateral transfers of military hardware complements the financial flows to third countries is the localization of physical power to enforce migration control those regions experiencing the most refugees.

Another way in which the externalization agenda is increasingly being pursued is through the privatization of migration (Gammeltoft-Hansen, 2011) reflecting the diverse ways in which the phenomenon of migration is increasingly being commodified. Thus, the Schengen Convention's codification of carrier sanctions in 1990 obliged transport companies to conduct pre-screening of travellers for visas, passports and other travel documents displaced the location of border control from Europe's territorial boundaries to airports across the globe. The carrier sanctions indirectly introduced private actors into the infrastructure of border control trading on their motivation to avoid substantial fines for facilitating irregular migration into Europe.

Yet, while border control may be an obstruction to the business interest of transport and travel agencies or the import of labour, it is also becoming a profitable emerging market for others. Thus, border control functions are increasingly outsourced to PSCs such as Finmeccanica and G4S. Some of these 'borderscape contracts' (Lemberg-Pedersen, 2012) pertain to the construction of control facilities in third countries while others concern the operation of control facilities.

For instance, the UK Border Agency [UKBA] contracted G4S to conduct all circulation of asylum seekers between holding centres in the UK via its Transport PLUS Service. The operation of the removal and holding centres was also outsourced to a number of PSCs with G4S managing centres like Brook House, Dungavel, Oakington and Tinsley House. Finally, G4S was also contracted by the UKBA to run all deportation flights out of Great Britain[1]. As such, G4S plays a vital role in the infrastructure of the UK borders. When it comes to the Finmeccanica contracts these illustrate how PSCs may be activated in favour of the externalization agenda. Thus, attempting to control the migration flows transiting Libya, the Commission and Frontex conducted technical missions to the country with a view to establishing cooperative agreements on migration control in 2004 and 2007. In 2008, moreover, a Friendship Pact was made where Italy agreed to pay Libya 255 million dollars annually over twenty years (Gazzini, 2009). Some of these funds were earmarked to the construction of border control infrastructure in Libya. This was outsourced to Italian companies, most notably Finmeccanica whose subsidiaries, like AugustaWestland, Alenia Aeronautica and SELEX Sistemi Integrati landed a range of contracts with Libya. Thus, AugustaWestland supplied Libya with 10 A109 Power helicopters for border control worth €80 million in 2006 and Alenia Aeronautica an ATR-42MP maritime patrol aircraft worth €31 million in 2008. Moreover, at the same time as the A109-deal Libya, Finmeccanica and AugustaWestland formed a joint venture called Libyan Italian Advanced Technology Company [LIATEC] to provide know-how, training, technology and equipment thereby to Libya's military and border infrastructures (Finmeccanica website). In 2009, Finmeccanica subsidiary SELEX Sistemi Integrati entered a €300 million contract for the construction of C3 (Command, Conquer and Communication) border infrastructure along Libya's borders with Chad and Niger (Lemberg-Pedersen, 2012).

The geostrategy of the networked border and its introduction of diffused forms of border-regulation can facilitate processes of neoliberalization (Peck and Tickell, 2002) and borderscape contracts shows how privatization of border control reconstructs the financial flows in borderscapes from public to private actors. Comparable to the bureaucratic crowd-out where foreign policy and transportation officials were replaced by internal security officials during the formulation of the Dublin Convention, national and supranational European structures tasked with border politics are thus increasingly experiencing a 'privatized crowd-out'. Importantly, the transfer of border activities from public to private hands does not signal the demise of borderscape governance. Rather, the planning and operation

1 Following the death of an Angolan national during a deportation in late 2010, G4S's contract was not renewed. This, however, did not put a halt to the outsourcing of deportations since the UKBA immediately contracted another PSC, Reliance Security Task Management Ltd. to perform this border function.

of borderscapes are displaced from formal to informal techniques of government through the introduction of new actors (see also Lemke, 2002, p. 11).

The reconfiguration of the power relations underpinning borderscapes through borderscape contracts illustrates how the transnational flows of funds in neoliberal processes can expand border control according to the agenda of containment. Taken together, the transnational flows of funds, personnel, equipment and competences, makes it relevant to examine the functionality of this decentralized border infrastructure more in depth.

Border-induced Displacement and the Sequencing of Forced Flows

A dominant description attached to EU border control has been that of 'Fortress Europe'. However, while this understanding fits well with some instances of border control, like the fences in Ceuta and Melilla, it does not capture the mobility going on within the borderscapes. Yet, the relation between Europe and Third World mobility in the European borderscapes is differentiated and the hypermobility of European border funds, equipment and personnel is crucial in bringing about the submobility of non-European migrants.

When we observe the coastal regions at the Atlantic and Mediterranean Seas it becomes clear that the very same islands and territorial waters which are specialized in luxury hotels and cruise-tours for European tourists are also the key infrastructural nodes of the networked European borderscape and veritable spider-webs for the surveillance, interception, detention and deportation of migrants from Africa and the Middle East. Most often, migrants reaching Europe territory through these regions have been intercepted numerous times and thus moved in and out of relations of inclusive exclusion. Border control should therefore not be understood only as a response to forced migration caused by other factors like conflict-induced displacement, development-induced displacement or environment-induced displacement, but also as a cause of displacement and forced migration in itself. Consequently, the humanitarian consequences of externalization should make us realize that forced migration is increasingly caused by border-induced displacement.

Conceptually speaking, the difference between original and border-induced forced migration is one between first order and second order displacement. Natural disasters, civil wars, foreign occupations or forced resettlement due to construction projects can displace people resulting in their forced migration. However, when these same people are intercepted, detained or deported in instances of border control they once again experience displacement, only this time induced by the inclusive exclusions of the European borderscape. Accordingly, border-induced displacement is not equivalent to the original displacement of people, which forced them to migrate, but instead functions as a second-order displacement imposed upon already displaced people. The displacement occurring in the EU's border system is therefore one where people are constantly being uprooted and transferred between control elements, thus entering a state of quasi-permanent displacement.

Due to the dynamics of diffusion, decentralization and restructuring in the European borderscapes, border-induced displacement is brought about by different actors and in different ways. One example is the Frontex-interceptions followed by the transportation of migrants to, say, the camps in Lampedusa, whereafter the migrants are further dispersed to camps on the Italian mainland. Another example is thus the common practice of both European and African states to transfer asylum seekers between camps and the Transport PLUS Service of G4S in Great Britain illustrate that both public police forces and private actors are involved in this activity. Moreover, as the case of Libya illustrate, cooperation between smugglers and police can also perpetuate border-induced displacement. As a 26-year old Eritrean, Daniel, explained Human Rights Watch:

> Every two or three days, the manager of Kufra camp took 25 or 30 persons at night and sold them to Libyan transporters so he could get money from us. Other people were just thrown in the desert. Sometimes they would take people in the desert and run over their legs with a car and just leave them. He sold me with a group of 25 or 30 people to a Libyan man who put us in a big house in Kufra and told us we needed to have our families send 200 dollars to pay for our release. (Human Rights Watch, 2009, p. 72)

According to Human Rights Watch smugglers also operate detention facilities and wear military uniforms giving migrants a hard time distinguishing between the informal and formal actors involved in their forced movement (Human Rights Watch, 2009, p. 75) since they are constantly transferred back and forth between them in trucks and containers.

Moreover, border-induced displacement also occur as a result of readmission agreements between states, for instance, Libya deported around 200,000 persons to other countries between 2003 and 2006 (Fortress Europe, 2007, p. 6). Following the networked rationality of externalization, however, the migrants' displacement often does not stop there as the countries they are deported to also have readmission agreements with other countries. The network of readmission agreements therefore form channels through which migrants are successively transferred from country to country illustrating the transnational character of the border-induced displacement created by externalization. Byrne and Shacknove have coined the apt term of 'chain deportations' to describe this forced traffic of migrants (Byrne and Shacknove, 1996, pp. 189–90). After the incidents at Ceuta and Melilla, for instance, several thousand migrants were rounded up and deported to other African countries, such as Algeria and Mauritania with which Morocco had readmission agreements. These in turn deported the migrants yet again and in the end many ended up in the countries from which they had originally fled (Costello, 2005, p. 47). Border-induced displacement is therefore a systemic feature of the European borderscapes and particularly obvious in the transnational buffer zone of externalized control.

Different from the walls of a fortress, the inclusive exclusions in the externalized European borderscapes do not pre-empt migration simply by halting it, but instead by transforming it into a different kind of mobility, namely border-induced displacement. Migrants are turned into various sequences of state-controlled forced flows. If we define a sequence as a series of element guided by a particular logic, the European borderscapes can be seen as relying on the construction of series of decentralized control elements, placing migrants in various inclusive exclusions designed to contain and circulate their mobility outside Europe. The production of these sequences of forced corporeal movement then depend on other sequences, of funds, personnel, equipment and information created by security apparatuses like financial instruments, ILOs, Frontex and the SIS and EURODAC databases and circulated within the control infrastructure. Externalization thus has an inherently dromopolitical dimension, that is, relying on politics of speed (Virillo, 1977). The faster the circulation of information between the control elements, the more rapid can instances of border control be directed against the 'threats' and 'risks' of migrants.

Seeing externalized border control as relying on a series of elements rather than specific control points, such as a specific wall, camps or patrol boats, highlights that it does not function as a wall accumulating migrants on the outside, but as an engine of serialized displacement. The border system thus enforce a controlled mobility where those intercepted are trafficked between its many elements through practices like Moroccan flight-deportations, European transports between camps, Italian–Libyan push backs or the day-long container transportations by the military, private militias or smugglers in Libya. Understanding the transnational European borderscapes as a series of elements guided by logic of the sequential production of forced flows may help us understand the dynamic and interrelated functionality of the EU's externalized control.

Conclusion

To conclude, as the ENP, Seville's Migration–development nexus and Frontex' HERA operations showed, externalization happens through multifaceted processes which may involve international negotiations, external governance or extra-territorialization. This made it possible to understand externalization as the construction of dynamic and fluctuating borderscapes outside Europe, guided by abstractions of knowledge and technologies, according with the EU agenda of containing migrants in regions of origin or transit. As a result of this, externalized borderscapes have both geopolitical and biopolitical aspects since they reterritorialize the geographic spaces of third countries in order to divide mobility into free and forced flows. Externalization is thus a flanking measure designed to safeguard the free circulation within Schengen by simultaneously excluding and subsuming migrants under the power of border control.

However, there were problems with the manner in which Agambenian biopolitics premised itself on the notion of sovereign power both because this reifies states' self-understanding as being capable of making exceptions out of migrants and also by conceptualizing them as passive instances of bare life. Yet, although many migrants in the externalized European borderscape experience exploitation and abuse, they also attempt to manoeuvre and resist the inclusive exclusions of border control by bribing border guards, seeking irregular work or demonstrating against their conditions showing that migrants are in possession of agency. Moreover, as the formulation of the Dublin Convention illustrated, state sovereignty needs to be differentiated in order to reflect both the bureaucratic and legal processes, which precipitate border politics as well as the multiplicity of formal and informal, national and supranational involved in the external European borderscapes.

Exploring the transnational character of externalization, the chapter went on to identify four instances of transnational flows, crucial to the construction of externalized border infrastructures, namely flows of funds, personnel, and equipment. These flows are crucial for the production of a power-geometry of European hypermobility and non-European submobility. It was argued that the existence of a fourth flow increasingly occurring when border control is transferred from public hand to PSCs creates further problems for underdetermined notions of sovereignty, since this flow blurs the divide between the public and private interests in borderscapes.

Finally, the chapter suggested that externalized border control should not only be seen as a reaction to displacement and forced migration, but as itself causing these phenomena. This was conceptualized through the notion of border-induced displacement. It was argued that this kind of displacement is not an incidental feature, but in fact the underlying logic behind externalization. Conceptualizing this logic as the sequencing of forced flows according to the containment agenda made it possible to focus on the interrelated functionality of this decentralized border infrastructure. At the end of the day, it was argued this provides a lens capable of assessing the transnational and systematic production of displaced and vulnerable existences in the European borderscapes.

References

Agamben, G. (1995), 'We Refugees', *Symposium*, 49(2): 114–19.

Agamben, G. (1998), *Homo Sacer: Sovereign Power and Bare Life*, Stanford, CA: Stanford University Press.

Agamben, G. (2005), *State of Exception*, Chicago, IL: The University of Chicago Press

Aeneas programme (2006), *Programme for Financial and Technological Assistance to Third Countries in the Area of Migration and Asylum. Overview of Projects Funded 2004–2006*, EuropeAid.

Balzacq, T. (2009), 'Frontiers of Governance: Understanding the External Dimension of EU Justice and Home Affairs', in T. Balzacq (ed.) *The External*

Dimension of EU Justice and Home Affairs: Governance, Neighbours, Security, Basingstoke: Palgrave Macmillan, pp. 1–34.

Betts, A. and Milner, J. (2007), *The Externalisation of EU Asylum Policy: The Position of African States*, Copenhagen: DIIS Brief December 2007.

Bigo, D. (1996), *Polices en reseaux: L'expérince européenne*, Paris: Presses de Sciences Po.

Bigo, D. (2000), 'Liaison Officers in Europe: New Officers in the European Security Field', in J.W.E Scheptycki (ed.), *Issues in Transnational Policing*, New York: Routledge, pp. 67–99.

Bigo, D. (2001), 'The Möbius Ribbon of Internal and External Securit(ies)', in M. Albert, D. Jacobsen and Y. Lapid (eds), *Identities, Borders, Orders. Rethinking International Relations Theory*, Minneapolis: University of Minnesota Press, pp. 91–116.

Bigo, D. (2007), 'Detention of Foreigners, States of Exception, and the Social Practices of Control', in Prem Kumar Rajaram and Carl Grundy-Warr (eds), *Borderscapes: Hidden Geographies and Politics at Territory's Edge*, Minneapolis, London: University of Minnesota Press, pp. 3–34.

Bakewell, O. (2008), '"Keeping Them in Their Place"': The Ambivalent Relationship between Development and Migration in Africa', *Third World Quarterly*, 29(7): 1341–58.

Boswell, C. (2003), 'The "External Dimension" of EU Immigration and Asylum Policy', *International Affairs* 79(3): 619–38.

Byrne, R. and Shacknove, A. (1996), 'The Safe Country Notion in European Asylum Law', *Harvard Human Rights Journal* 9: 185–226.

Chimni, B.S. (1998), 'The Geopolitics of Refugee Studies', *Journal of Refugee Studies* 11(4): 350–74

Costello, C. (2005), 'The Asylum Procedures Directive and the Proliferation of Safe Country Practices: Deterrence, Deflection and the Dismantling of International Protection?', *European Journal of Migration and Law* 7: 35–69.

Council of the European Union (29.10.2002) *Seville European Council 21 and 22 June 2002: Precidency Conclusions*, OR 13463/02, POLGEN 52.

Council of the European Union (2005), *A Strategy for the External Dimension of JHA – Global Freedom, Security and Justice*: 2. Note from Presidency to Council, 14366/3/05 REV 3.

Dean, M. (1999), *Governmentality: Power and Rule in Modern Society*, London: Sage Publications Ltd.

Dean, M. (2002), 'Liberal Government and Authoritarianism', *Economy and Society*, 31(1): 37–61.

Deleuze, G. and Guattari, F. (1987), *A Thousand Plateaus: Capitalism and Schizophrenia*, London: Athlone Press.

Dietrich, H. (2004), 'The Desert Front – EU Refugee Camps in North Africa?', *Konkret*, 12.

Doctors without Borders, (2006), *Activity Report Marokko 2006*. Available at: http://www.msf.org/msfinternational/invoke.cfm?objectid=959FC5B0-5056-

AA77-6CF9DC7DBEBF276A&component=toolkit.article&method=full_html [accessed 12.5.2010].

Doctors without Borders (2009), *Migrants' Living Conditions in Lesvos Detention Center Spark Discontent*. Available at: http://doctorswithoutborders.org/news/article.cfm?id=4021&cat=field-news [accessed 12.5.2010].

Finmeccanica:http://www.finmeccanica.it/Corporate/EN/Corporate/Press_and_Media/Comunicati_stampa/args/detail/details~press~Comunicati_Stampa~2006~press_dettaglio_00396.shtml///index.sdo [accessed 21.6.2011].

Fortress Europe (2007), *Escape from Tripoli: Reports on the Conditions of Migrants in Transit in Libya*, Rome: Fortress Europe.

Foucault, M. (2007), *Security, Territory, Population*, Basingstoke: Palgrave Macmillan.

Foucault, M. (2008), *The Birth of Biopolitics: Lectures at the Collége de France, 1978–1979*, Basingstoke, New York: Palgrave Macmillan.

Frontex, http://www.frontex.europa.eu/examples_of_accomplished_operati/art5.html [accessed 22.1.2010].

Gammeltoft-Hansen, T. (2011), *Access to Asylum*, Cambridge: Cambridge University Press.

Gazzini, C. (2009), 'Assessing Italy's Grande Gesto to Libya', *Middle East Report*, March 16.

Geddes, A. (2000), *Immigration and European Integration. Towards Fortress Europe?*, Manchester: Manchester University Press.

Gibney, M. (2007), 'Forced Migration, Engineered Regionalism and Justice between States', in S. Kneeborne and F. Rawlings-Sanaei (eds), *New Regionalism and Asylum Seekers: Challenges Ahead*, Oxford: Berghahn Books.

Goldschmidt, E. (2006), 'Storming the Fences: Morocco and Europe's Anti-Migration Policy', *Middle East Report* 239, Summer.

Guild, E. (2006), 'The Europeanisation of Europe's Asylum Policy', *International Journal of Refugee Law*, 18(3–4): 630–51.

Hyndman, J. (2000), *Managing Displacement: Refugees and the Politics of Humanitarianism*, Minneapolis, London: University of Minnesota Press.

Human Rights Watch (2009), *Pushed Back, Pushed Around*. Available at: http://www.hrw.org/en/reports/2009/09/21/pushed-back-pushed-around [accessed 12.5.2010].

Lavenex, S. and Ucarer, E.U. (2004), 'The External Dimension of Europeanization: The Case of Immigration Policies', *Cooperation and Conflict*, 39(4): 417–45.

Lemberg-Pedersen, M. (2012), 'Private Security Companies and the European Borderscapes', in T. Gammeltoft-Hansen and N.N. Sørensen (eds), *The Migration Industry: The Commercialization of International Migration*, London: Routledge.

Lemke, T. (2002), 'Foucault, Governmentality, and Critique', *Rethinking Marxism*, 14(3): 172–9.

Lemke, T. (2005), 'A Zone of Indistinction – A Critique of Giorgio Agamben's Concept of Biopolitics', *Outlines. Critical Social Studies*, 7(1): 3–13.

Lucht, H. (2011), *Darkness before Daybreak: African Migrants Living on the Margins in Southern Italy Today*, Berkeley: University of California Press.

Massey, D. (1993), 'Power-Geometry and a Progressive Sense of Place', in J. Bird, B. Curtis, T. Putnam, G. Robertson and L. Tickner (eds), *Mapping the Futures: Local Cultures, Global Change*, New York: Routledge: 59-69.

Official Journal of the European Union (2004), *Council Regulation (EC) No 377/2004 of February 2004 on the creation of an immigration liaison officers network*. L 64: 1–2.

Paludan, A. (1981), 'Refugees in Europe', *International Migration Review*, 15(1–2): 69–73.

Peck, J. and Tickell, A. (2002), 'Neoliberalizing Space', *Antipode*, 34(3): 380–404.

Rancière, J. (2004), 'Who is the Subject of the Rights of Man?', *South Atlantic Quarterly* 103(2–3): 297–310.

Rajaram, P.K. and Grundy-Warr, C. (eds) (2007), *Borderscapes: Hidden Geographies and Politics at Territory's Edge*, Minneapolis, London: University of Minnesota Press.

SIPRI (2009), *Arms Transfers to Central, North and West Africa*, SIPRI Background Paper.

SIPRI (2010), *Trends in International Arms Transfers*, SIPRI Fact Sheet.

UNHCR Global Report (2001), Donor Profiles. Available at: file:///Users/mlpedersen/Desktop/EURECO-konference/Financial%20flows/UNHCR%20donations/UNHCR%20Donor%20Profiles%202001.webarchive [accessed 3.9.2010].

UNHCR Global Report (2004), Donor Profiles. Available at: file:///Users/mlpedersen/Desktop/EURECO-konference/Financial%20flows/UNHCR%20donations/UNHCR%20Donor%20Profiles%202004.webarchive [accessed 3.9. 2010].

UNHCR Global Report (2009), Donor Profiles. Available at: file:///Users/mlpedersen/Desktop/EURECO-konference/Financial%20flows/UNHCR%20donations/UNHCR%20Donor%20Profiles%202009.webarchive [accessed 3.9. 2010].

Van Munster, R. (2009), *Securitizing Immigration: The Politics of Risk in the EU*, Basingstoke: Palgrave Macmillan.

Virillo, P. (1977), *Speed and Politics*, Los Angeles: Semiotext(e).

Walters, W. (2004), 'The Frontiers of the European Union: A Geostrategic Perspective', *Geopolitics*, 9(3): 674–98.

Chapter 3
Mobile Regions: Competitive Regional Concepts (not only) in the Danish–German Border Region

Martin Klatt

Europe of Regions – Fiction or Reality?

Regions are 'in' in national, European and international politics. Regionalization, regional integration, the 'Europe of Regions', regional identity, regional development – the term region occupies many buzz words in contemporary politics and social science discourses, particularly since the 1990's worldwide *New Regionalism* discourse (Beeson, 2005; Calleya, 2000; Katz, 2000; Keating, 1998, 2004; Lahiri, 2001). Europe is rescaling (see Keating, Chapter 1, in this volume) and seemingly moving away from the clear cut nation state system of the first half of the century. Regional concepts of territory experience a revival in many European states, often transcending state borders. Still, in spite of the multiplying use of the term 'region' in politics and social science discourses, there is still a lack of clear definitions, when we use the term region, acknowledging that regions' status within the hybrid systems of governance and identity prevailing in today's globalized world are different. This is even more the case in relation to borders and border studies, where there is a lack of clarity with the terms of border- and cross-border region and their relation to regional identity. On the basis of the Danish–German border region of Sønderjylland–Schleswig, this article will present different regional concepts applied to this region in different periods of history and reflect on the flexibility of 'regions' as well as characteristics of successful region-building.

I will start by looking at regions as a geographical space. In 1995, the Council of Europe (CoE) proposed the following definition for a cross-border region: 'A potential region, inherent in geography, history, ecology, ethnic groups, economic possibilities and so on, but disrupted by the sovereignty of the governments ruling on each side of the frontier' (CoE 1995: 10). Thus generalized from (cross-)border region, a region should be inherent in geography, history, ecology, ethnic groups and the economic fabric. Here, inherence is central: regions as the lowest form of inherent space, using different categories to measure inherence.

Anssi Paasi has gone further by defining regions as historically contingent structures, whose institutionalization is based on their territorial, symbolic and institutional shaping, and regional identity being an abstraction that can be used to analyse the links between social actors and the institutionalization process of a region (Paasi, 1986, 2002). Region is linked to territoriality, but while all territories are regions, not all regions are territories – a territory differs from a region in that its boundaries and the resources therein are under the control of people (Paasi, 2009). I would argue further that regional concepts through agents strive for territorialization and that the degree of territorialization is a decisive parameter for the continuous success and stability of regional concepts, which will be demonstrated below. The same is valid for identity. Following Paasi's definition above, regional identity requires a regional institutionalization process. This process, though, does not have to be limited to political institutionalization and can extend to several other dimensions. Identity should here be understood both as self-identification and outside perception, meaning that regional identity needs both a self-identification of people within the region, as well as an outside identification of a specific identity of people within that region.

Extended to a cross-border perspective, on one side we witness highly flexible definitions of cross-border regions by agents involved in cross-border cooperation, often depending on the definition of legibility of EU funding schemes (Perkmann, 2007). While the above mentioned CoE definition of a cross-border region reflects on potential regions indicating their non-existence, social science definitions of cross-border regions are restricted to define them as unions of sub-national regions with a potential to integrate into a common, cross-border space of identification and action (Schmitt-Egner, 1998, 2001) or to develop a certain degree of policy entrepreneurship (Leresch and Saez, 2006; Perkmann, 2002, 2007). Many empirical case studies have estimated their ability to create a degree of familiarity or belonging beyond an elite of transnational borderlanders (Martinez, 1994) or regionauts (Löfgren, 2008; O'Dell, 2003) as low (Andersen, 2004; Bucken-Knapp, 2001; Dupeyron, 2009; Ehlers, 2001; Heddebaut, 2001; B. Henningsen, 2011; Jukarainen, 2003; Keil and Löfgren, 2011; Klatt and Kühl, 2008; Klatt and Herrmann, 2011; Liikanen, 2008; Strüver, 2005; Yndigegn, 2011).

This article qualifies the success of regional concepts by their ability to create a feeling of belonging, of *Heimat* contrasting to its traditional opposite, *Fremde*. The concept of Heimat is here understood as the core feeling of belonging for each individual. Heimat is the familiar, family and personal heritage. It is a place an individual feels at home and to which she/he has a distinct relationship. Human beings are genetically disposed to a spatial concept of home in contrast to the foreign, and this disposition is the background for a socially constructed Heimat-territory, which may go beyond the usual space, human beings physically cover in their daily lives (Bastian, 1995). Especially in Germany, Heimat has since the late 19th century been socially constructed as an interchangeable representation of the local, the regional and the national community (Confino, 1993), also in Schleswig, as will be demonstrated below. Finally, Heimat is primarily defined in cultural and emotional categories as area, scenery, dialect, familiarity, only secondarily in social categories as family relations, friends and social activities (Bastian, 1995, p. 86).

Hence, this article demonstrates how people internalized the abstract concepts of regions and became part of an *imagined community* (Anderson, 1991), as well as why and how especially the regional concept of Schleswig-Holstein was able to produce familiarity and a stable regional identity, while others have remained elite concepts never reaching a significant familiarity (Spierings and van der Velde, 2008) among social actors within the region.

The Danish–German Border Zone of Sønderjylland/Schleswig as an Example for Competitive Concepts of Regions

Today's land border between Denmark and Germany on the Cimbrian peninsula was drawn in 1920 after two plebiscites, which were an integral part of the World War I peace settlements between the Allies and the Central Powers (Fink, 1979; Goldstein, 2002). It divided the former Duchy of Schleswig, which had been the scene of a century old national conflict with two wars and two national assimilation projects fought out on its territory. The division after a plebiscite should assure a just border according to the right of national self-determination, and a voter turnout around 90 per cent with rather clear results (75 per cent for Denmark north of the border, 80 per cent for Germany south of the border) indicate the success of the attempt to draw a border as just as possible in a nationally mixed border zone. Nonetheless, the new border had severe economic consequences for the region (Andersen, 2008), and since post-World War II European integration and Denmark's accession to the EC there have been several attempts to overcome the socially and economically divisive character of the border, culminating in the establishment of a Euroregion in 1997 (Andersen and Thönnes, 2006; Klatt, 2005, 2006a, 2011; Klatt and Herrmann, 2011). Evaluations show, though, that the Euroregion has not succeeded in effective cross-border governance, stable regional institutions or a conscious regional identity among the people living there (Andersen, 2004; Hjalager, 2004, 2009; Klatt, 2006a, 2011).

To analyse why the attempts to recreate a cross-border region have been so difficult so far as well as how the concept of a cross-border region interrelates with other regional concepts, I will in a first step return to the mentioned CoE definition of a region (CoE, 1995, p. 10). To apply this definition on the Danish–German border region, we have to check, whether it is possible to set up quantifiable criteria for inherence for this specific bounded space, and, in consequence, how this bounded space actually can be delineated. This implies the danger of returning to the concept of 'natural borders' that has caused many violent conflicts on national borders during the 19th and 20th centuries. On the other hand, it is obvious that the border drawing of 1920 also in the Danish–German case divided a region, which used to be:

Coherent in geography/ecology

The Cimbrian peninsula is characterized by three sets of landscapes. On the North Sea coast, the unique Wadden Sea goes over into rich coastal marshes. In the midlands, poor, sandy soil dominates, while the East Coast is characterized by fertile moraine soil providing rich farming opportunities. Thus, geographically we here have an East–West divide, but North–South inherence. The ecology influenced the geography with sea faring being characteristic for the coastal areas, cattle breeding dominating the marshes, rich farmhouses, large scale trading and early industry on the Eastern moraine soil and poor, subsistence farming with little trading in the midlands.

Coherent in history

Here, of course, we are on soft grounds. Historians have written history to justify national projects and thus construct nations and align them with territories long into the 20th century, and are still doing so today especially in newly independent nations, sub-state nations and regions striving for secession. For the Sønderjylland/Schleswig case, I have earlier demonstrated, how problematic it is to use history to justify contemporary political programs of reunification or recreation of a 'historic' cross-border region (Klatt, 2006b, 2010). Nonetheless, it can be ascertained that the border line drawn in 1920 was new and did not reflect historic borders in the region as a whole. Though, it follows rather closely the line dividing the use of German and Danish language in church services in rural parishes (Henningsen, 2009, p. 19).

Coherent in ethnic groups

Archaeological and historic evidence suggest a relatively high mobility in the region, mainly on a north–south route and using the sea routes. Two north–south highways are still called 'oxen way' and 'army way', indicating the drift of cattle to southern markets, and armies passing through. Hedeby/Haithabu on the Schlei inlet was a large trading city since the 9th century. Place names indicate a Jutian/Danish population in the areas northeast of the Eider River, while Frisians settled the Western islands and marshes during the 9–10th centuries. Since then, though, mobility has mixed the population, and there is evidence for both Slavic and Saxon immigration since medieval times. So, ethnic coherence has not been static.

Economic fabric – economic opportunities

The economic fabric and economic opportunities have been characterized by the ecology/geography of the region. Here, again, an East–West divide has been characteristic and is still visible today; reinforced by the structure of infrastructure: Main highways and all functioning rail lines are north–south. The large harbours are on the East Coast reflecting a historically higher trade volume there. The poor midlands have improved agriculturally on either side of today's border, but still lag behind the more fertile soils on the East and West Coast. Industrialization has almost exclusively been restricted to the East Coast, while tourism is strong on the West Coast, especially on the German side of today's border.

As I have demonstrated here, geographical coherence does not reflect the political spatialization of the Danish–German border region. On the contrary, as for most territories in Europe, spatialization is the result of historical processes of power and spatial control. Exactly delimited territories are a phenomenon of the modern state system evolving since the 17th century, often characterized as the Westphalian state system.

Early Concepts of Territory and Domain in Schleswig

The 'Ritterschaft' – an Early Bearer of Regionalism?

Territorialization and consolidation of the Duchy of Schleswig and county of Holstein (elevated to Duchy in 1474) began in the 10th and 11th century, when the Duchy was carved out as a fief of the then still rather feeble Danish kingdom (Riis 1998). In the 1460 treaty of Ribe, the nobility of both territories elected Danish king Christian I. as duke and count, receiving special privileges in return. These privileges continued to support some territorial autonomy, as well as co-determination of politics and taxes by the nobility united in the so-called Schleswig-Holstein equestry or knighthood ('Ritterschaft'), a cooperation still existing today. Contrary to the nobility in the kingdom of Denmark, the Ritterschaft prevailed in its privileges, and consequently absolutism was never introduced in the Duchies. Until the 19th century struggle, the Ribe privileges played a central role as specific 'Landesrechte' against Danish efforts to centralization and administrative homogeneity. The Ritterschaft had a decisive function as bearer of power and identity. It is difficult to say, though, how far this disseminated into the population beyond the nobility.

17th-century Sovereign Gottorf

During the 15th and 16th centuries, three partitions of Schleswig were implemented to satisfy dynastic needs. As a result, a minor line of the Danish royal family resided at the Gottorf court in Schleswig-city, ruling a part of Schleswig-Holstein, the Duchy Schleswig-Holstein-Gottorf, with a complicated set up of territorial specialities including joint rule of some territories with the king of Denmark. Thus, when territorially rounded-up, more centralized states became the rule in Europe, we face a rag-rug of different administrative entities in what are today Southern Jutland and the German state (*Land*) Schleswig-Holstein (Schlaber 2007).

In the course of the Danish–Swedish conflict following the 30 Years War, Gottorf was released from its fiefdom and gained sovereignty from the kingdom Denmark with the peace treaty of 1658. In the following years, an impressive baroque court was set up at Schleswig's Gottorf castle, and a university was established by Duke Christian Albrecht in Kiel in 1664. Though difficult to assess in their impact on regional identity, we can assume that this institutional build-up would have had longer lasting effects had the era not been troubled by repetitious conflict and war. Denmark did not respect the treaty of 1658 in the long run and strove to reintegrate Schleswig into the domain of the kingdom, eventually abolishing all kinds of special privileges and administrative differences (Henningsen, 1998). Formal reintegration was achieved in 1721 after the Great Nordic War. Nonetheless, integration did not mean administrative harmony, as historically developed special regimes continued, in fact beyond the integration into Prussia, into the end of the 19th century (Schlaber, 2007). Here it was again the strong position of the local nobility, the Schleswig-Holstein Ritterschaft, who obstructed administrative harmony on retaining their privileges.

Early Identity Construction

It is difficult to ascertain from when we could reasonably detect 'regional identity' or a regional consciousness in the Duchies of Schleswig and Holstein. The existence of pre-national, regional identities here has not been researched yet, while a similar pre-national, ethnic based Danish identity can be ascertained for the territories becoming core Denmark since medieval times (Hansen, 2005, p. 93). Having in mind the image of a patchwork rug, when looking at maps visualizing the partitions of Schleswig and Holstein in the 16th century, there is little indication of a concept of region. Different territories seem to be connected only coincidentally, as a result of dynastic, inheritances and matrimonial arrangements. Different estates with different rights and privileges show a typical feudal map, far from a streamlined coherent administration. In the 17th century, war with foreign soldiers passing through created further chaos and hardships on the local populations. On the other side, one of the longest lasting historic popular narratives recorded in many contemporary village chronicles relates to the 'Polackenkrieg' referring to Polish soldiers plundering and marauding in the region during the many wars of the 17th century (Lange 2003, p. 244).

From the 17th century, a personal cult celebrating the Danish king became popular (for the following paragraph: Henningsen 2000). The king visited the Duchies regularly to inspect military installations and the administration, as well as to respect social relations to family members residing there. The king's visits were a popular attraction. Beggars, men as well as children, and young girls flocked around the king's carriage, and the monarch generously gave to the poor and others entertaining him. The king usually dined in public, so all his subjects could enjoy their monarch 'live'. Guards of parade were established in most cities of the Duchy in honour of the monarch. In the 1730's, the king's birthday was celebrated at the governor's residence in Schleswig with a dinner and ball. In 1750, a celebration of the king's birthday with a festive service is recorded from Husum church. In 1760, absolutism's 100th anniversary in Denmark was celebrated with big festivities all over the Duchy. Thus, by 1760, the Danish king had been established as 'land-father' in all Schleswig, and the population's loyalty to him seems to be a fact (Hansen et al., 2008b, p. 389; Henningsen, 1998). Here, though, we witness the development of a pre-national identity rather than regional, as it was the clear task of the crown to assemble the former patchwork rug into a rather homogeneous empire – aligning with similar trends under enlightened absolutism all over Europe. Later national and/or ethnic criteria were not important for this early identity construction.

The task of administrative homogeneity could not be accomplished in the Duchies, though, as regional opposition delayed or hindered many reform projects , which were implemented in the core kingdom of Denmark in the 18th century. Schleswig and Holstein developed a distinct administrative history as the 'German provinces' of the Danish conglomerate state, administered by the 'German Chancellery' in Copenhagen, which since 1721 was led by a renowned representative of the nobility of the Duchies. Likewise, naming of the population as 'Schleswigians – Schleswiger' or 'Holsteinians – Holsteiner' became synonymous, with a domination of the use of 'Holsteiner' for people from both Duchies also in official texts - as for example the Indigenate Law of 1776 reserving public offices for 'Danish, Norwegians and Holsteinians' (Frandsen, 2008; Hansen et al., 2008b, p. 447; Henningsen and Hansen, 1997). Newer historic research also indicates Holstein's disintegration from the Holy Roman Empire approaching the Danish kingdom beyond the imperial border, to a much stronger effect than in other peripheral regions of the empire (Mörke, 2000).

Contrary to the 17th, the 18th century brought a long period of peace and thus economic and commercial development to the Duchies. This period was brought to an end with the Napoleonic wars, were Denmark came into a very difficult situation due to Napoleon's activities in northern Germany. British insecurity about Danish neutrality led to demands to deliver the Danish fleet. When rejected, the British bombed Copenhagen in 1807 and captured the fleet. French pressure at the southern border of Holstein by the annexation of Hamburg and Lübeck into the French empire then ended Danish neutrality by an alliance with France. The bombing of Copenhagen and capture of the Danish fleet aroused a wave of patriotism also measurable in Schleswig and Holstein through public demonstrations and collections for the victims. Thus, around 1800 there were many witnesses of a conglomerate state patriotism in multicultural Schleswig-Holstein–Denmark–Norway, before nationalism started to occupy people's self-identification from the mid of the 19th century (Fangel et al., 1984; Japsen, 1979).

The Long 19th Century – Region- and Nation-Building

Regional Turns National

While conglomerate state patriotism and the king as land-father still were flourishing in the Duchies at the turn of the 18th/19th century, a conflict between Danish and German evolved at about the same time in the conglomerate state's capital, Copenhagen. It was rooted in distrust between the central administration and the Schleswig-Holstein nobility, the so-called 'Ritterschaft'. While the central administration tried to harmonize administrative practices and state policies, the Ritterschaft cautiously guarded its inherent privileges. A more popular expression of the conflict was visible during the Struensee crisis, when German physician Johann Friedrich Struensee became royal physician to the mentally ill king Christian VII., who made him de facto regent in 1770. At the same time, Struensee had a love affair with the queen and probably fathered her daughter. He was dismissed in a coup by reactionary forces and beheaded in 1772. Popular resentments against him played on his German heritage and the dominance of Germans at the Copenhagen court, in administration, commerce and at the university. The Struensee interregnum and its aftermath went into history as the German feud. It resulted in, amongst others, the introduction of Danish as command language in the military and a general commitment to strengthen the Danish language within the conglomerate state, but also in a general negative attitude against all 'German' by the Copenhagen bourgeoisie (Feldbæk and Winge, 1991). A counterpart indicating early nationalization can be detected during the temporary incorporation of Holstein into Denmark 1806–1815, which caused reactions in the Duchy indicating a regional identity expressed by the refusal of the prospect of becoming Danish (Frandsen, 2008, p. 16).

By the mid-19th century, three regional concepts were competing with the traditional conglomerate state patriotism in the Duchies: Schleswig-Holsteinism, a conglomerate state oriented Schleswigianism and a new, Eider-Danish Schleswigianism. Schleswig-Holsteinism as a frame of identity evolved from the University at Kiel. It was closely related to contemporary dreams to establish a democratic German empire after the defeat of Napoleon. In a speech at the university's celebration of the battle of Waterloo on 7 July 1815, professor of history Friedrich Christoph Dahlmann referred to the German people, united in spite of the division into many bigger and smaller states, and claimed Schleswig to belong to it, too, because of the Duchy's long connection with Holstein. At the same time, cities in Schleswig as well as the nobility addressed the king to establish estate assemblies in both Duchies, or even a united Schleswig-Holstein assembly. In 1830, Uwe Jens Lornsen, civil servant on the Frisian island of Sylt, agitated for a democratic constitution for Schleswig-Holstein, which should have resulted in political autonomy of the Duchies. Lornsen's project was welcomed and supported by many liberal intellectuals and citizens in the Duchies. While Lornsen was imprisoned, the king granted the Duchies estate assemblies in 1831.

Election to these pre-parliaments was limited to some 2 per cent of the population, nonetheless they became a forum for liberals, who turned more and more national during the 1840's, dividing especially the Schleswig assembly into two factions: Schleswig-Holsteinians and Eider-Danish. The first sought to incorporate a united Schleswig-Holstein into the German Confederation, while the latter sought to only incorporate the Duchy of Schleswig into the Danish kingdom.

How far did these concepts of region and identity disseminate into the population? Personal identification is difficult to ascertain in general, especially regarding 'regional identity' vs. other identities. Sources we have are usually restricted to specific groups in the population. This is especially the case in the 18th and 19th century, where sources expressing feelings of identity are limited to few, literate people, whom we would classify as intellectuals today. On agent level, interpretation of action can be multi-contextual. For example, we can assume that the corps of civil servants in Schleswig's opposition to strategies strengthening the role of the Danish language in the Duchy's administration – did express a different identity than their counterparts within the Danish kingdom. It is difficult to classify, though, whether we here witness a culturally defined German identity, or a regional Schleswig-Holstein identity, a multinational conglomerate state identity or simply conservatism opposing changes of hitherto practices.

During the period described as the national breakdown, the 1840's, several activities were used to mobilize the population indicating identity formation and identity shift. Music (singing festivals) and petitions played an important role on both sides. So, while Hans Schultz Hansen does not believe that a Schleswig-Holstein identity went beyond the administrative elite of the Duchies before the 1830s–40s (Hansen et al., 2008a, p. 16), in 1844 petitions circulated in the Duchies protesting against Danish national agitation for the total integration of Schleswig into the kingdom and thus separating it from Holstein. Here, locals especially in the Eastern parishes of Schleswig expressed a distinctive Schleswig identity wishing to be connected to the 'brethren' in Holstein and likewise to the 'brethren' in Denmark. A 'people's address' to the king in 1847 signed by almost 4,000 inhabitants mainly in Northern Schleswig confirmed this attitude (Hansen, 1998). The core of the Danish-Schleswigian movement was independent farmers, followed by tenants ('Husmænd') and craftsmen. Very few were estate owners, estate managers, owners of large farms, civil servants or had a liberal profession. Geographically, it was concentrated in the northern and eastern parts of Schleswig.

The German-Schleswig-Holsteinian movement was supported by the society's 'elite' and people striving to connect with it, and it was strongest in the towns. Here, educated bourgeois like physicians, lawyers and civil servants, as well as ship-owners, merchants and industrialists took part in the national activities. In the rural areas, it was predominantly estate owners, estate managers and owners of large farms, who chose the Schleswig-Holstein movement by signing petitions and joining associations. But the large share of the lower class population did not take part visibly in the national competition in its early stage (Hansen et al., 2008a, p. 86).

The conflict culminated in the revolution of 1848, almost simultaneously resulting in a national-liberal Eider-Danish government in Copenhagen and a secessionist revolutionary government in Kiel. The following war of independence (1849–1851) ended with the Schleswig-Holstein insurgents' defeat and a return to the status quo ante. Danish language policies in the interwar years did not appease the conflict, but rather further alienated loyal Schleswigians, who were not eager to give up their German cultural and linguistic affiliation. After the Danish–Prussian–Austrian war of 1864, Schleswig and Holstein were annexed by Prussia and incorporated as a Prussian province in 1867. This was not necessarily what the revolutionaries in 1848 had been fighting for. Even though they aimed at the integration of Schleswig-Holstein into a greater Germany, they hoped for becoming a federal state of this greater Germany to maintain their clearly defined regional identity (Fangel et al., 1996). Nonetheless, regional self-identification became secondary to the rising glory of the German empire eventually after the German victory over France in 1870 and the wedding in 1881 of the Schleswig-Holstein princess Auguste Viktoria to later emperor William II. The new authorities also did their best in successfully incorporating Schleswig-Holstein's population into the Prussian state (Jahnke, 2005; Lange, 2003). In North Schleswig, the Danish population experienced hardships of political and cultural suppression culminating in the late 1880's and early 1890's, the so-called Köller period. Nevertheless, there existed a Danish cultural and political life outside the official German institutions, born by committed idealists from the region and supported by non-governmental organizations in Denmark. Here, though, regional characteristics were the exception. The bonds to Denmark were clearly national: Danishness, the Danish language, Danish songs and Danish culture in general (Fink, 1999).

The Schleswig-question was put on the international agenda again after the defeat of Germany in World War I. The peace treaty included provisions for a plebiscite to decide, how much of the former Duchy should return to Denmark (Fink, 1979). This plebiscite can well be considered the ultimate act in the nationalization of regional identity, as the people in the region were to choose between Germany and Denmark, with no third, regional alternative. Nonetheless, regional and *Heimat* elements played a certain role in the propaganda preceding the plebiscites of 10 February and 14 March 1920. Here, especially the German side was able to successfully connect the regional concept of Schleswig-Holstein with German nationalism (Hoffmann et al. 2001; Jebsen and Klatt 2011; Pohl 1996). Voting turnout was above 90 % reflecting the deep emotional involvement of the population in this question. Thus, since 1920, a border legitimized by the people has separated Denmark and Germany. The regional concept of Schleswig was ended by the division of its territory and its supporters were forced into the national minorities remaining on either side. These, though until the present day are defined nationally as German North Schleswigians and Danish South Schleswigians. Both minorities kept the term 'Schleswig' in their self-identification and nomenclature. Although today, they see themselves as cross-border agents, culturally bridging between Denmark and Germany and recreating a cross-border space (Malloy 2010).

Schleswig-Holstein – A Stable Concept in Different Times

After 1920, only one regional concept proved to be sustainable in the Danish–German border region: Schleswig-Holstein. Its administrative status as a province of Prussia during the Weimar Republic and a 'Gau' in the Nazi party organization secured some institutional stability. This institutional stability became even stronger after WW II, when the allies decided to dissolve Prussia. Instead of creating a North German *Land* in the British zone, the military administration decided to establish four: Schleswig-Holstein and Niedersachsen, and the city-states of Hamburg and Bremen. Contemporary politicians already realized the problematic aspect of a dominantly agricultural state of Schleswig-Holstein burdened with refugees from Germany's lost eastern territories, but until today different approaches to restructure the federal structure of Northern Germany have not come beyond expert discussions. So, Schleswig-Holstein became just another, if not the foremost example on the characteristic continuation of German nationhood as a patchwork of regions and states, the setup of Germany as a conglomeration of regional states, and the concept of a nation composed of regional identities, and thus the locality, Heimat and region as the cradle of German nationhood (Confino, 1993).

While the regional concept of Schleswig-Holstein had a clear national, German, anti-Danish alignment in the 19th and most of the 20th century (Andresen, 2010), late 20th and 21st century Schleswig-Holstein politicians have constructed a specific Nordic identity including her Danish heritage and her connection with the Baltic and Nordic countries (Henningsen, 2011; Sinner, 2011). Danishness, or a new-Scandinavianism has become quite popular in Schleswig-Holstein since the 1970's, reflecting images of liberality, relaxedness, modern pedagogics and a more just society; not the least reflected in the increasing political success of the Danish minority's political party at state and regional elections (Klatt and Kühl, 2006; Kühl, 2011). The same party, the South Schleswig Voters Association (Südschleswigscher Wählerverband, SSW), has had no scruples using this support, transforming its historic South Danish narrative of South Schleswig as an integral part of Danish history in contrast to the German Holstein into a new Schleswig-Holstein 'Für uns im Norden' (for us in the North) slogan since the 2008 municipal elections, and has made the continued existence of the present federal state of Schleswig-Holstein in her territorial 1937 borders (which are quite different from her borders in 1920, 1867, 1773 and 1460) as a precondition for supporting a state government after the 2012 elections (SSW, 2011). Today, even the historic personal union of Schleswig-Holstein and Denmark is presented as a model for the European Union and its motto Unity in Diversity (Hill, 2010). The regional concept of Schleswig-Holstein thus has demonstrated an interesting adaptive flexibility securing its survival in often short-lived political agendas shaped by different *Zeitgeist*.

Sønderjylland – A Dying Regional Concept?

'Sønderjylland' (=Southern Jutland) can be called the Danish counter concept to Schleswig-Holstein. Relating to the early medieval differentiation of Northern and Southern Jutland with the Kongeå creek as their border, it is synonymous with the later Duchy of Schleswig bordered at the Eider-Levensau to Holstein. Sønderjylland was revived as territorial name, when national conflict began in the early 19th century, as the 'original Danish' name for the Duchy of Schleswig, where Schleswig was associated with Germanization (Henningsen and Hansen, 1997). Contrary to Schleswig, Sønderjylland clearly associates to the territory's Danish-ness, its Danish heritage in language and popular culture (Eskildsen, 1936). Sønderjylland focused strongly in the resistance against Germanization from 1864–1920 but had a far less articulated role in the plebiscite agitation than 'Schleswig-Holstein' (Jebsen and Klatt, 2011). After its 're-unification' with Denmark in 1920 every effort was made to integrate the northern part of Sønderjylland into Denmark, both economically and socially (Andersen, 2006, 2008), and a regional 'Schleswigianism' remained an option exclusively identified with the German minority, eventually associated with a return of the region to Nazi Germany (Hansen, 2001; Werther, 2010). When recreated as Sønderjyllands Amt (Southern Jutland County) with the administrative reform in 1970, the regional concept of Sønderjylland had a revival, not the least as political agent in a cross-border perspective and founding father of the Euroregion Sønderjylland-Schleswig. Considerable popular opposition against the Euroregion demonstrated the national context of Sønderjylland as Denmark's border region to Germany, though, where the issue of national sovereignty clearly superseded independent regional agency and identity (Damsgaard, 1999; Klatt, 2006b, 2006a). With the administrative reform in 2007, counties were abolished and Sønderjylland became a part of the new Region South Denmark. Interestingly enough, only the German minority in Denmark voiced strong criticism against this administrative reform. In 2009, its party campaigned on 'vote Southern Jutian' for the municipal elections resulting in the first increase in votes since the 1980's.

Sønderjylland-Schleswig: Re-enacting a Region?

In 1997, Southern Jutland County (Sønderjyllands Amt), the counties of Schleswig-Flensburg and Nordfriesland and the city of Flensburg established Region Sønderjylland-Schleswig as a cross-border, Danish-German Euroregion. Contrary to the development in other European border regions, the establishment of a Euroregion here caused a very emotional debate in Denmark (Damsgaard,

1999; Klatt, 2006a). The region started, nonetheless, and has initiated several cross-border initiatives. It's most successful probably is the establishment of a cross-border information centre for commuters and others. Evaluations have criticized the Euroregion, though, for falling short of its original ambitions to create a common, cross-border action space, recreating the cross-border region of Schleswig (Hjalager, 2004, 2009; Klatt and Herrmann, 2011). Especially its political institutions were criticized for being ineffective, resulting in an organizational restructuring in autumn 2011, when the twice annual regional assembly was abolished to be replaced by an annual stakeholder conference. The daily operations are now led by a board representing the administrative members and the two national minorities in Sønderjylland-Schleswig.

Especially the idea of reuniting the German regional Schleswig(-Holstein) identity and the regional Southern Jutian identity into a cross-border Schleswig-Sønderjylland identity proved to be a failure. 15 years after the violent clashes on whether to establish a cross-border region, pragmatic cooperation exists between agents in many different fields. Also cultural cooperation is quite successful, but research among youths does not indicate trends to common identity. On the contrary, clear images of the 'other' exist (Yndigegn et al., 2002; Yndigegn, 2003), and a recent project idea to invite border region youths ('generation Schleswig') to discuss their idea of the German–Danish border region totally failed to attract participants (Flensborg Avis, 22 June 2011). With the Danish administrative reform, cross-border region building has focused on other geographical spaces. This is partly induced by the extension of the Interreg-eligible area from Interreg III to IV, which has 'created' new regional concepts as the 'Growth Region South Denmark-Schleswig-KERN' (2011). In my opinion it remains highly doubtful, whether a four year Interreg project aimed at business cooperation can manifest a region and regional identity as understood in our working definition (following Paasi, 2002).

Conclusion – Heimat, Region or Nation?

Regions are socially constructed spaces. This becomes evident, when analysing and comparing the different concepts of region applied to the Danish–German land border region on the Cimbrian peninsula. While nationalization in Denmark strengthened an existing process of increasing centralization, in Germany, nationalization coincided with the strengthening of regional identities (Frandsen, 2008, p. 52). The concept of Schleswig-Holstein is a good example on how a strong regional identity has been aligned with national identity. It proved to be endurable from late medieval, early modernity, absolutism, liberalism, the imperial Kaiserreich, Nazism to the democratic Federal Republic of Germany of today. This is an impressive example of a symbiosis of regional and national identity. Especially, as territorial continuity was only partial, but historiography did their part to confirm the teleological development to the present day land of Schleswig-Holstein (Lorenzen-Schmidt, 2010). Thus today, Schleswig-Holstein unifies the local *Heimat*, the north German region as well as the German nation and even a Nordic-European dimension of identity in an age of high mobility.

References

Andersen, K. and Thönnes, F. (2006), 'Abschlussbericht. Dänisch-deutsche Arbeitsgruppe zur Förderung der grenzüberschreitenden Mobilität', København/Bonn: Beskæftigelsesministeriet/Bundesministerium für Arbeit und Soziales.

Andersen, M.L. (2004), *Grænsen i hverdagen – grænsen i hovedet*, in I. Adriansen and K.M. Pedersen (eds) *Skrifter fra Museet på Sønderborg Slot*, Sønderborg/Aabenraa: Museet på Sønderborg Slot/Institut for Grænseregionsforskning.

Andersen, M. (2006), 'Nation-Building and Border Identities: The Economic Reconstruction of the Schleswig Border Region, 1919–1924', in M. Hurd (ed.), *Borderland Identities. Territory and Belonging in North, Central and East Europe*, Eslöv: Gondolin: Baltic and East-European Studies, pp. 225–54.

Andersen, M. (2008), *Den følte grænse. Slesvigs deling og genopbygning 1918–1933*, Aabenraa: Skrifter udgivet af Historisk Samfund for Sønderjylland.

Anderson, B. (1991), *Imagined Communities*, London/New York: Routledge.

Andresen, K. (2010), *Schleswig-Holsteins Identitäten. Die Geschichtspolitik des Schleswig-Holsteinischen Heimatbundes 1947–2005*, Quellen und Forschungen zur Geschichte Schleswig-Holsteins, Neumünster: Wachholtz.

Bastian, A. (1995), 'Der Heimat-Begriff. Eine begriffsgeschichtliche Untersuchung in verschiedenen Funktionsbereichen der deutschen Sprache', in H. Henne, H. Sitta, and H.E. Wiegand (eds) *Reihe Germanische Linguistik*, Tübingen: Max Niemeyer Verlag.

Beeson, M. (2005), 'Rethinking Regionalism: Europe and East Asia in a Comparative Historical Perspective ', *Journal of European Public Policy* 12: 969–85.

Bucken-Knapp, G. (2001), 'Just a Train-ride Away, but Still Worlds Apart: Prospects for the Øresund Region as a Binational City', *GeoJournal* 54(1): 51–60.

Calleya, S.C. (ed.) (2000), *Regionalism in the Post-Cold War World*, Aldershot: Ashgate.

CoE (1995), 'Manuel de coopération transfrontaliére à l'usage des collectivités locales et régionales en Europe', Strasbourg: CoE.

Confino, A. (1993), 'The Nation as a Local Metaphor: Heimat, National Memory and the German Empire, 1871–1918', *History and Memory* 5(1): 42–86.

Damsgaard, J.I. (1999), 'Der må være en grænse – osse mellem gode naboer. En analyse af modstanden mod oprettelsen af det grænseoverskridende samarbejdsorgan Euroregion Slesvig', University of Aarhus.

Dupeyron, B. (2009), 'How History Becomes a Cross-Border Matter: Death Foretold of a French–German–Swiss Textbook', *Curriculum Journal* 20(4): 345–64.

Ehlers, N. (2001), 'The Utopia of the Binational City', *GeoJournal* 54(1): 21–32.

Eskildsen, C. (1936), *Dansk Grænselære*, Copenhagen: Reitzel.

Fangel, H. et al. (eds) (1984), *Der nationale Gegensatz – de nationale modsætninger 1800–1864* Quellen zur Geschichte der deutsch-dänischen Grenzregion; Flensburg/Aabenraa: Institut für Regionale Forschung und Information im Deutschen Grenzverein e.V./Institut for Grænseregionsforskning.

Fangel, H. et al. (eds) (1996), *Der nationale Gegensatz – De nationale modsætninger* (Quellen zur Geschichte der deutsch-dänischen Grenzregion; Flensburg/Aabenraa: Institut für Regionale Forschung und Information im Deutschen Grenzverein e.V./Institut for sønderjysk Lokalhistorie.

Feldbæk, O. and Winge, V. (1991), 'Tyskerfejden 1789–1790. Den første nationale konfrontation', in O. Feldbæk (ed.), *Dansk Identitetshistorie* 2, København: C.A. Reitzels Forlah, pp. 9–109.

Fink, T. (1979), *Da Sønderjylland blev delt,* 3 vols, Aabenraa: Institut for Grænseregionsforskning.

Fink, T. (1999), *Båndene båndt. Forbindelsen over Kongeåen 1864–1914*, 2 vols, Aabenraa: Institut for Grænseregionsforskning.

Frandsen, S.B. (2008), *Holsten i helstaten. Hertugdømmet inden for og uden for det danske monarki i første halvdel af 1800-tallet*, København: Museum Tusculanums Forlag.

Goldstein, E. (2002), *The First World War Peace Settlements 1919–1925*, Seminar Studies in History, London: Pearson Education.

Hansen, H.S. (1998), 'Schleswigsche Identität 1840–1848', in H.B. Christensen (ed.), *Geschichte Schleswigs. Vom frühen Mittelalter bis 1920*, Aabenraa: Institut for Grænseregionsforskning, pp. 127–36.

Hansen, H.S. (2001), 'Mindretal og flertal i Nordslesvig omkring 1940', in R. Bohn, U. Danker, and J. Kühl (eds), *Nationale mindretal i det dansk-tyske grænseland 1933–1945*, Aabenraa: Institut for Grænseregionsforskning, pp. 122–41.

Hansen, H.S. (2005), *Hjemmetyskheden i Nordslesvig 1840–1867 – den slesvig-holstenske bevægelse. Bind I: 1840–1850*, 2 vols, Skrifter udgivet af Historisk Samfund for Sønderjylland, Aabenraa: Historisk Samfund for Sønderjylland, vol. 1.

Hansen, H.S., Henningsen, L.N., and Porskrog Rasmussen, C. (eds) (2008a), *Sønderjyllands Historie 2. Efter 1815*, 2 vols, Aabenraa: Historisk Samfund for Sønderjylland, vol. 2.

Hansen, H.S., Henningsen, L.N., and Porskrog Rasmussen, C. (eds) (2008b), *Sønderjyllands Historie 1. Indtil 1815* 2 vols, Aabenraa: Historisk Samfund for Sønderjylland, vol. 1.

Heddebaut, O. (2001), 'The Binational cities of Dover and Calais and their Region', *GeoJournal* 54(1): 61–71.

Henningsen, B. (2011), 'On Identity – No Identity. An Essay on the Constructions, Possibilities and Necessities for Understanding a European Macro Region: The Baltic Sea', available on http://www.bsr2011.eu/2011/10/21/is-there-a-baltic-sea-region-identity/ [accessed 21.11.2011].

Henningsen, L.N. (1998), 'Die Zusammenführung Schleswigs unter dänischer Herrschaft – Idee oder Realität?', in H.B. Christensen (ed.), *Geschichte Schleswigs. Vom frühen Mittelalter bis 1920*, Aabenraa: Institut for Grænseregionsforskning, pp. 127–38.

Henningsen, L.N. (2000), *Kongen kommer! Kongebesøg og kongeloyalitet i Sydslesvig*, Grænseforeningens Årbog, København: Grænseforeningen.

Henningsen, L.N. (ed.) (2009), *Sydslesvigs danske historie*, Studieafdelingens udgivelser, nr. 62, 2. reviderede udgave. Flensborg: Studieafdelingen ved Dansk Centralbibliotek for Sydslesvig.

Henningsen, L.N. and Hansen, H.S. (1997), '"Sønderjylland" og "Schleswig"', *Sønderjyske Årbøger*, pp. 5–26.

Hill, T. (2010), '"Vielfalt in der Einheit – Schleswig-Holstein und Dänemark als Vorbild für Europa?" Eine Unterrichtseinheit zu Landesgeschichte und europäischem Geschichtsbewusstsein', *Demokratische Geschichte* 21: 231–53.

Hjalager, A.-M. (2004), 'Evaluierung der Arbeit in der Region Sønderjylland-Schleswig 2000–2004', Aarhus: Advance/1.

Hjalager, A.-M. (2009), 'Evalueringen 2009 af Region Sønderjylland-Schleswig', Aarhus: Advance1.

Hoffmann, E. et al. (eds) (2001), *Der nationale Gegensatz- De nationale modsætninger 1914–1933*, Quellen zur Geschichte der deutsch-dänischen Grenzregion, Aabenraa/Flensburg: Institut for Grænseregionsforskning/ Institut für schleswig-holsteinische Zeit- und Regionalgeschichte.

Jahnke, C. (2005), 'Die Borussifizierung des schleswig-holsteinischen Geschichtsbewusstseins, 1866–1889', *Zeitschrift der Gesellschaft für Schleswig-Holsteinische Geschichte* 130: 161–90.

Japsen, G. (1979), 'Statspatriotisme og nationalfølelse i Sønderjylland før 1848', *Historie* 13(1–2): 107–22.

Jebsen, N. and Klatt, M. (2011), 'The Negotiation of National, Regional and Minority Identity During the Plebiscites Following the First World War', *The Association for the Study of Nationalities 2011 World Convention: Charting the Nation Between State and Society*, New York City.

Jukarainen, P. (2003), 'Definitely Not the End of Nations – Northern Borderlands Youth in Defence of National Identity', *Youth: Nordic Journal of Youth Research* 11(3): 217–34.

Katz, B. (ed.) (2000), *Reflections on Regionalism*, Washington, DC: Brookings Institution Press.

Keating, M. (1998), *The New Regionalism in Western Europe*, Cheltenham: Edward Elgar.

Keating, M. (2004), *Regions and Regionalism in Europe*, The International Library of Comparative Policy, Cheltenham: Edward Elgar.

Keil, D. and Löfgren, K. (2011), 'Grenzenlose Visionen. Die Fehmarnbeltregion im Transitkorridor Hamburg-Kopenhagen', *NORDEUROPAforum* 21(1): 9–39.

Klatt, M. (2005), 'The Danish-German Border Region as an Example for Integration in the European Context', in J. Kühl and M. Weller (eds), *Minority Policy in Action: The Bonn-Copenhagen Declarations in a European Context*, Aabenraa: Institut for Grænseregionsforskning, pp. 139–60.

Klatt, M. (2006a), *Fra modspil til medspil?: grænseoverskridende samarbejde i Sønderjylland-Schleswig 1945–2005*, Aabenraa: Institut for Grænseregionsforskning.

Klatt, M. (2006b), 'Common, Cross-Border Regional History as an Approach to People-to-People Cooperation and Cross-Border Regional Integration', in M. Hurd (ed.), *Borderland Identities. Territory and Belonging in Central, North and East Europe*, Baltic and East European Studies, Eslöv: Gondolin, pp. 109–46.

Klatt, M. (2010), 'Re-enacting a Region – Why is it so Difficult to Revise Border Changes?', in E. Medijainen and O. Merkelsmann (eds), *Border Changes in 20th Century Europe*, Münster: Lit Verlag, pp. 21–30.

Klatt, M. (2011), 'Grenzüberschreitende politische Zusammenarbeit in der deutsch-dänischen Grenzregion', in J. Beck and B. Wassenberg (eds), *Grenzüberschreitende Zusammenarbeit leben und erforschen (2): Governance in deutschen Grenzregionen. Beiträge aus dem Forschungszyklus zur grenzüberschreitenden Zusammenarbeit der Universität Strassburg und des Euro-Institutes*, Studies on the History of European Integration, Stuttgart: Franz Steiner Verlag, vol. 2, pp. 269–88.

Klatt, M. and Kühl, J. (2006), *SSW – Minderheiten und Regionalpartei in Schleswig-Holstein 1945–2005*, Flensburg: Studieafdelingen ved Dansk Centralbibliotek for Sydslesvig.

Klatt, M. and Kühl, J. (2008), 'National Minorities and Crossborder Cooperation between Hungary and Croatia. A Case Study of Baranya, Hungary, and Osijecko-baranjska County, Croatia', in A. Bloed, R. Hofmann, J. Marko, J. Mayall and M. Weller (eds), *European Yearbook of Minority Issues*, 6, Leiden/Boston: Martin Nijhoff Publishers, pp. 193–210.

Klatt, M. and Herrmann, H. (2011), 'Half Empty or Half Full? Over 30 Years of Regional Cross-Border Cooperation within the EU: Experiences at the Dutch-German and Danish-German Border', *Journal of Borderlands Studies* 26(1): 65–87.

Kühl, J. (2011), 'Von der Abgrenzung zum Miteinander 1955–2010', in L.N. Henningsen (ed.), *Zwischen Grenzkonflikt und Grenzfrieden. Die dänische Minderheit in Schleswig-Holstein in Geschichte und Gegenwart*, Flensburg: Studieafdelingen ved Dansk Centralbibliotek for Sydslesvig.

Lahiri, S. (ed.) (2001), *Regionalism and Globalization*, London: Routledge.

Lange, U. (ed.) (2003), *Geschichte Schleswig-Holsteins. Von den Anfängen bis zur Gegenwart*, Neumünster: Wachholtz Verlag.

Leresch, J.P. and Saez, G. (2006), 'Political Frontier Regimes: Towards Cross-Border Governance?', in M. Perkmann and N.-L. Sum (eds), *Globalization, Regionalization and Cross-Border Regions*, Basingstoke/New York: Palgrave Macmillan, pp. 77–99.

Liikanen, I. (2008), 'New Neighbourhood and Cross-Border Region Building: Identity Politics of CBC on the Finnish-Russian Border', *Journal of Borderlands Studies* 23(3): 19–39.

Löfgren, O. (2008), 'Regionauts: the Transformation of Cross-Border Regions in Scandinavia', *European Urban and Regional Studies* 15(3): 195–209.

Lorenzen-Schmidt, K.-J. (2010), 'Paradigmenwechsel in der schleswig-holsteinischen Landesgeschichtsschreibung und Geschichtsvermittlung 1945–2009', *Demokratische Geschichte* 21: 139–56.

Malloy, T. (2010), 'Creating New Spaces for Politics? The Role of National Minorities in Building Capacity of Cross-Border Regions', *Regional & Federal Studies* 20(3): 335–51.

Martinez, O.J. (1994), *Border People. Life and Society in the U.S.-Mexico Borderlands*, Tucson: University of Arizona Press.

Mörke, O. (2000), 'Holstein und Schwedisch-Pommern im Alten Reich. Integrationsmuster und politische Identitäten in Grenzregionen', in N. Jörn and M. North (eds) *Die Integration des südlichen Ostseeraums in das Alte Reich*, Quellen und Forschungen zur höchsten Gerichtsbarkeit im Alten Reich, Köln: Böhlau, pp. 425–72.

O'Dell, T. (2003), 'Øresund and the Regionauts', *European Studies: A Journal of European Culture, History and Politics* 19: 31–53.

Perkmann, M. (2002), 'Euroregions: Institutional Entrepreneurship in the European Union', in M. Perkmann and N-L. Sum (eds), *Globalization, Regionalization and Cross-Border Regions*, Basingstoke/New York: Palgrave Macmillan, pp. 103–24.

Perkmann, M. (2007), 'Policy Entrepreneurship and Multilevel Governance: A Comparative Study of European Cross-Border Regions', *Environment and Planning C-Government and Policy* 25(6): 861–79.

Pohl, K.H. (1996), 'Demokratisches Schleswig-Holstein? Zur politischen Kultur bei den Volksabstimmungen 1920', *Demokratische Geschichte* 10: 105–24.

Paasi, A. (1986), 'The Institutionalization of Regions: A Theoretical Framework for Understanding the Emergence of Regions and the Constitution of Regional Identity', *Fennia* 164(1): 105–46.

Paasi, A. (2002), 'Bounded Spaces in the Mobile World: Deconstructing "Regional Identity"', *Tijdschrift voor Economische en Sociale Geografie* (*Journal of Economic & Social Geography*) 93(2): 137–48.

Paasi, A. (2009), 'The Resurgence of the "Region" and "Regional Identity": Theoretical Perspectives and Empirical Observations on Regional Dynamics in Europe', *Review of International Studies* 35 (Supp.): 121–46.

Region South Denmark-Schleswig, 'Growth South Denmark-Schleswig-KERN region', available at http://www.schleswig-holstein.de/cae/servlet/contentblob/390978/publicationFile/zusammenWachsenDE.pdf [accessed 12.12.2011].

Riis, T. (1998), 'Wann entstand das Herzogtum Schleswig?', in H.B. Christensen (ed.), *Geschichte Schleswigs. Vom frühen Mittelalter bis 1920*, Aabenraa: Institut for Grænseregionsforskning, pp. 7–10.

Schlaber, G.L. (2007), *Hertugdømmet Slesvigs forvaltning. Administrative strukturer og retspleje mellem Ejderen og Kongeåen ca. 1460–1864*, in L.N. Henningsen (ed.), *Skrifter udgivet af Studieafdelingen ved Dansk Centralbibliotek for Sydslesvig*, Flensburg: Studieafdelingen ved Dansk Centralbibliotek for Sydslesvig.

Schmitt-Egner, P. (1998), 'Grenzüberschreitende Zusammenarbeit' in Europa als Gegenstand wissenschaftlicher Forschung und Strategie transnationaler Praxis. Anmerkungen zur Theorie, Empirie und Praxis des Transnationalen Regionalismus', in G. Brunn and P. Schmitt-Egner (eds), *Grenzüberschreitende Zusammenarbeit in Europa. Theorie – Empirie – Praxis*, Baden Baden: Nomos.

Schmitt-Egner, P. (2001), '"Transnationaler Regionalismus" als Gegenstand der Politikwissenschaft', in J. Bellers and C. Rosenthal (eds), *Die gesellschaftliche Basis von Aussenpolitik. Internationale Wirtschaft, Umwelt, Ideologien, Regional- und Entwicklungspolitik*, Münster: Lit-Verlag, pp. 386–441.

Sinner, K. (2011), *Schleswig-Holstein – das nördliche Bundesland. Räumliche Verortung als kulturelles Identitätskonstrukt*, Kieler Studien zur Volkskunde und Kulturgeschichte, Münster: Waxmann.

Spierings, B. and van der Velde, M. (2008), 'Shopping, Borders and Unfamiliarity: Consumer Mobility in Europe', *Journal for Economic and Social Geography* 99(4): 497–505.

SSW, available on www.ssw.de [accessed 21.12.2011].

Strüver, A. (2005), 'Bor(der)ing Stories. Spaces of Absence along the Dutch-German Border', in H. van Houtum, O. Kramsch, and W. Zierhofer (eds), *B/ordering Space*, Aldershot: Ashgate, pp. 207–22.

Werther, S. (2010), 'Nazi, German and Danish Identities in Danish North Schleswig, 1932–1938', in M. Hurd (ed.), *Bordering the Baltic. Scandinavian Boundary-Drawing Processes 1900–2000,* Münster: LIT, pp. 69–104.

Yndigegn, C. (2003), *Unge og regional identitet. Forventninger og indstilling til livsbetingelser og livsmuligheder i den dansk-tyske grænseregion*, Aabenraa: Institut for Grænseregionsforskning.

Yndigegn, C. (2011), 'National Borders, European Integration, Baltic Cooperation and Supra-National Spatial Identities', in M. Hurd (ed.), *Bordering the Baltic. Scandinavian Boundary-Drawing Processes, 1900–2000*, Münster: Lit-Verlag, pp. 127–49.

Yndigegn, C., Pedersen, K.M., and Schack, M. (2002), *Unges holdninger til nabolandet–enforundersøgelse*,Aabenraa: InstitutforGrænseregionsforskning.

PART II
Researching Everyday Practices at Borders

Chapter 4
Revisiting the Anthropology of Policy and Borders in Europe

Thomas M. Wilson

It is often easier to identify ways in which scholars, who come from all social science and humanities disciplines, differ in their theoretical and methodological approaches to borders, boundaries and frontiers than it is to identify that which draws them together as practitioners of border studies. Most, if not all border scholars are aware that international borders act simultaneously as bridges and barriers to movement, communication and understanding. It is clear that all border studies involve in some measure the recognition and analysis of territory (including space and place) and people (including identities, culture, and society). But movement and mobility, bridges and barriers, territory and identity, are just starting points in a very wide range of perspectives on how geography and humans intersect at the margins of other social, political, economic and cultural entities.

Given this range in possible approaches, perhaps I should not have been surprised by the response I received, when at the conference, which provides the basis for the volume in which this essay is included, I asked the assembled participants about what if anything might we agree which ties us together as border studies scholars. I asked this by also acknowledging the many professional and national barriers to agreement, as well as the constraints put on us all by our varied disciplinary backgrounds. The response was underwhelming: very few conference participants took me up on my offer. Yet there *are* many things which draw us together in border studies, and which made the conference itself interesting on many levels to many people, or so I concluded based on my observations and discussions over the days following my own presentation. And there are many common themes, tropes, motifs, subjects and objects that are woven throughout border studies, to a degree that at the very least resulted in the establishment of a professional association of scholars of border studies with wide membership particularly in Europe and North America.

In this essay I seek to examine briefly a few of the themes which I conclude involve most if not all of us in border studies. I do this by exploring how my own academic discipline, social anthropology, has contributed and will continue to contribute to border studies in Europe. In the first section I discuss how various ways of looking at space and history demonstrate that, far from there being a borderless Europe, borders are still very much with us in the member states of the European Union (EU), particularly in the guise of re-bordering and de-bordering, which often are processes that shift the physical manifestations of state borders to different spaces. Borrowing from the insightful critique of a so-called borderless world recently offered by Liam O'Dowd (2010), I argue that history must be included in all of our approaches to borders because not only does it prevent crude assertions and typologies about the uniqueness of contemporary globalization and neoliberalism, but it also allows us to see the relations between levels of political and social integration within and across state borders, in localities and regions, in much sharper relief.

In the second section of the essay I examine the anthropology of borders through a consideration of its approaches to public policy. Here too a historical dimension is necessary, for the absence of historical context undermines much that is offered as new and ground breaking in the anthropology of current policy. A combination of approaches in the anthropology of borders, which investigates the intersections of space, history, politics and power, can invigorate the anthropology of policy in Europe, which in turn will give anthropology more opportunity to contribute to public debate on the transformations in the borders and frontiers of Europe.

Border Studies and Border Theory

State borders are the principal sites and contexts to border studies, however they are defined. In almost all *border studies*, international borders provide the object if not the subject of analysis. Given the fact that sovereignty, territory, citizenship, national identity and so much more are inextricably linked to national borders, and despite the simple premise that all of these are constructed, contingent and compromised concepts with diverse meanings and practices, national and state borders are self-evidently central to border studies.

The metaphors of borders and frontiers also figure prominently in contemporary scholarly analyses of individual and group identities, and as part of such analyses the related metaphors of cultural flows are often utilized to describe relations between people, in what is often styled as the politics of identity. The boundaries which divide and align us in terms of identities and cultures are very often formulated in scholarly terms through the metaphorical use of geopolitical borders. In this *border theory*, we cross borders to identify with others; we build borders to keep them out. These metaphors operate on almost a universal basis, as may be witnessed by the multiplicity of uses in the term 'borders' when discussing the creative movement of individuals and groups to transcend a social boundary that they or others associate with them. Yet all of these metaphors find their potency and relevance because of the master narratives and monumental significances of state borders in the lives of so many people of the planet. Thus, the themes in border theory cannot be easily divorced from the flows, obstructions and stoppages which occur within the geopolitical and geo-economic spaces of our individual and collective public lives, in the very real worlds of society, economy and politics.

The irony of having borders as centrepieces within the analysis and rhetoric of a borderless new world order of globalization has not escaped the attention of scholars. The contradictions inherent in using borders as important motifs in scholarship that seeks to demonstrate how borders no longer matter as much as they once did is especially apparent in globalization rhetoric. This rhetoric often relies on assertions that nation-states and their respective borders have been weakened if not made impotent due to new movements and innovations in technology, capital, information and people. This oft-asserted borderless world has seemingly occurred while the number of states and their borders has proliferated, states have increased their capacities to provide both security and surveillance at borders, new technologies have made states more invasive and effective in gathering information on citizens and visitors than at any time in the past, and so-called neoliberalism has made the state much more of a public actor in all aspects of everyday life than ever before (for reviews of some of these contradictions regarding borders in a borderless world, see Balibar, 2009; Mann, 2007; O'Dowd, 2010). Much of this examination of the changing importance and definition of international borders has given rise to a focus on 'border work', i.e., the processes of neo-bordering, re-bordering, and de-bordering that displace or replace the functions long associated with states at their borders (Rumford, 2008; van Houtum, 2005; O'Dowd, 2010). These contradictions, wherein we must find out how other institutions and agencies handle the jobs of states at borders, or handle the functions of borders in geographical areas distant from the territorial limits of a state, also highlight, not surprisingly, the boundary between border theory and border studies, a boundary marked by many convergences and divergences (Heyman, 1994).

In summary, border theory refers to cultural studies and other scholarly approaches to identity and who we are and where we belong. Borrowing from the glossary of studies of geopolitical borders, i.e., those prescribed in law, on maps, with institutional sanction, border theory seeks to show how identity and other aspects of cultural life act as barriers or bridges between and among people, with particular emphasis on ethnic, racial and gender identity. Border studies on the other hand focuses on the geopolitical borders that seem to have given their metaphorical meaning to border theory, i.e., border studies concentrate on administrative, political and legal borders as they relate to all manner of political, social, cultural and economic matters. Of course, one may easily do border theory work within a context of border studies, and border studies may easily inform border theory. They are not mutually exclusive. But the more post-modern aspects of border theory have certainly become popular and salient in some scholarly circles, and rarely are border studies trumpeted as being particularly theoretical. This relative absence of critical social theory in much border studies scholarship might be seen as a negative common factor, but one way to make the connection more positive would be to reassess borders from the perspective of history.

Borders and History

Recent critiques of border studies have focused on the relative dearth of 'historical reflexivity', wherein scholars often avoid the historical positioning both of the borders they study and of border studies itself (O'Dowd, 2010, p. 1032). This myopia regarding the histories that matter to scholars, to their disciplines, to their mutual national professional and intellectual traditions, and to the borders and borderlanders under scrutiny become particularly salient when examining the state borders that frame most if not all border studies. As O'Dowd (2010, pp. 1031–2) points out, most views of international borders that are contained within border studies are steeped in the ideal of national state borders, rather than of borders situated in space and time. Historical, empirical and comparative studies of state borders, O'Dowd argues, may very well subvert the ideal constructions of state, nation, sovereignty and citizenship which have been so important to both border studies and to what I have termed above border theory.

This lack of historical depth and reflexivity has led to exaggerated claims in border studies. These concern the importance and uniqueness of contemporary border transformations; an inability to recognize the past in the present and to see borders as aspects in the *longue durée* of state and imperial formations; and the failure to see contemporary state borders as being similar to and different from other sorts of geopolitical borders (O'Dowd, 2010, p. 1032). Thus O'Dowd calls for a 'recalibration' (2010, p. 1033) of border studies so as to right the balance between space and history in order to focus on the boundedness of state borders, as a variable that should be understood historically. Europe is often privileged in this ahistorical approach to national borders in a rush to theorize the EU as the exemplar of how a new sort of polity can recreate economy, society and culture in a relatively borderless world of sovereign but interdependent states.

One thing that emerges from such critiques is that the consideration of space and groups of people in border studies must be tempered with a consideration of their histories. History and space intersect the framing, construction and practices of power and the politics of identity and culture in borderlands. Moreover, the historical reflexivity which inspires so much of the critique of the national state and its borders must also be applied to other geopolitical borders, such as those of regions and localities. While these latter sorts of borders owe a great deal to both the real and ideal of state borders, together with state and now supranational borders in Europe they give meaning and substance to the transformations that seem to vitalize so much of meaning in the daily lives of Europeans, and so much that impels the proliferation in border studies that is the concern of this essay and others in the collection to which it contributes. It is clear that state borders inform all of border studies, but it should be equally clear – and here social anthropology has played a role – that all border studies also rely heavily on other border motifs, experiences and examples, on other forms of boundedness and boundaries, whether they be of provinces, regions, communes, cities, villages or ecosystems. These other borders are most often subnational in character, i.e., they exist at levels below that of the national state, but in Europe levels above the state, namely that of the EU, must also be included in the wider historical reflexivity that energizes so much in border studies. One lens through which contemporary and historical borders may be viewed, because of its relevance to all social sciences and to most if not all Europeans, is that of policy, a subject that has increasingly interested social anthropologists at and across all types of European borders.

Borders and the Anthropology of Europe

History has long been a key feature of the anthropology of Europe, which has also long focused on various types of borders in regional, state, national and imperial formations. Both the Boasians in North America and many social anthropologists working in the pre- and post-World War II period throughout the Anglophone world saw anthropology and history as complementary. Eric Wolf (for example, in Wolf, 1982) combined ethnography with history in order to investigate the political economy of culture, capitalism and imperialism. Wolf and John Cole's research in northern Italy (Cole and Wolf, 1974) was in fact the principal founding model in anthropology for borderland and border regional research in Europe. They, along with other students of Wolf in the 1970s (see, for example, Schneider and Schneider, 1976; Hansen, 1977), were fostering a new anthropology of regions, in order to provide a way forward from past community studies, to allow anthropologists to study and critique the nation and state.

As Cole remembered in his assessment of European anthropology (1977), borrowing from Schneider, Schneider and Hansen (1972), a region is not just an expression of geography, or a culture area in the sense of a collection of communities with a shared cultural tradition; rather it is a unit of political ecology, where local resources and people are organized by elites who are interposed between communities within the region and the wider nation and state. In his view and that of other Wolfians, community, regional, national and international levels of integration were in dynamic feedback relations with each other. And what better places to see these dialectical relations at work, in localities with daily and direct relevance to region, nation and state, than in international borderlands?

Today, a 'borderless Europe' is better conceived in my view as a 're-bordered Europe' and 're-bordering Europe'. The anthropological and ethnographic examination of these many new and transformed borders, including the effects on those borders that seem to have been removed or seem to be more permeable than they once were, is a continuing and important need in anthropology in Europe, precisely because of the tens of millions of Europeans who live and work at international and regional borders, and who define themselves in terms of identity and cultural practice, as borderlanders, people of the borders, perhaps even borderers. Their identification with borders is often instrumental, but is often also affective. It is precisely because borders are so important in their lives, and are so important to local, regional, national and European projects, that in my view any anthropology that defines its aspirations in ways that relegate the day to day roles of borders in peoples' lives to the dustbin of some old-fashioned sort of anthropology is a profession and scholarly discipline in trouble.

It would seem on the surface at least that anthropology as a profession would find it hard to distance itself from border studies in Europe. After all, it is patently obvious in quotidian life everywhere on the continent that borders have rarely mattered more. In Europe over the last generation there has been an increase in the numbers of states and borders, new forces of supranational and international integration, and new and continuing conflicts over sovereignty and internal and external national security. The media and governments also play a role in the way borders work by daily bringing the issues of borders and war, migration, illicit economies, and national and regional identities to our computers, televisions and smart phones.

However, although it would appear that anthropology has much to investigate, describe and analyse ethnographically at all sorts of borders, recent popular directions in professional anthropology have presented new hurdles to the continuing interest by anthropologist in local, regional, state and other boundaries. Anthropology as an academic discipline has to a great extent moved away from the concerns which I argue should still be central to an anthropology of borders in Europe. This development in anthropology reflects this conclusion reached by George Marcus (2007, pp. 32, 39–40), one of the leading cultural anthropologists over the last twenty-five years who is largely responsible for the redirection in ethnography towards reflexivity and away from the modernistic authority of cause and effect research and writing:

> change, politics, activism, post-colonialism, globalization – now are the milieu in which most career-making research projects are defined … there is for the foreseeable future no turning back to the archival function of ethnography in the old 'peoples and places' domain of inquiry; the bodies of theory, practices and topics which established this interest are in abeyance; in some instances played out; or simply not able to be pursued because the conditions in many places under which fieldwork must be undertaken will not permit it. (Marcus, 1997).

As Marcus makes clear, it is not fieldwork conditions that are the main culprit in the change in ethnographic practice, or that theories regarding politics in places like borderlands are played out. Rather, it is professional anthropology today which will not allow much that is labelled as the 'old' domain of ethnography to be pursued. However, despite the popularity in certain anthropological circles of the position articulated by Marcus, many ethnographers continue to study peoples in and of places. This is especially true of anthropologists who do border studies, and particularly those who study public policy in borderlands.

This interpretive turn in ethnography, which we can date to the mid-1980s, has led to a transformation of anthropology in the US, which is still home to the greatest number of anthropologists in the world, and where it is often difficult today to find PhD dissertations in cultural anthropology that contain any data other than interview excerpts or any attempts to establish cause and effect relationships. Any observations are purely tangential or accidental, or at best part of a cultural background to allow the interviewer to construct and assess the interview. What is missing is hard data about population, their activities at work and at leisure, their political and economic contexts, and group beliefs, values and attitudes. These are missing precisely because anthropologists have eschewed the more traditional approaches to community and culture, due to a widespread acceptance that these are no longer tenable as analytical or descriptive devices. Many anthropologists persist in this despite the overwhelming evidence that most people in the world use some form of community as an identifier of who they are and where they belong – here community should at the least be approached as an emic category – and culture still operates as the best way to express much about the meanings we attribute to things in our social and individual worlds.

In my view, anthropological efforts to find and theorize social permeability and cultural dissonance, to ask questions that need the researcher to move between field locations that do not constitute a solitary 'site', today often have little to do with the changing dimensions and bounded nature of culture and society, and more to do with the changing nature of professional ethnography. In these days of ethnographic conversation-making, where much that passes for scholarly reflexivity is little more than the recorded conversations with respondents, and where much ethnographic data can most kindly be labelled as culturally sensitive interviewing, the focus of ethnography is reduced to that of the network of respondents that the researchers construct around themselves.

However, this rather soft approach to ethnography is neither universal or without its own problems. The major trends in method and theory in much social and cultural anthropology today still come up squarely against the transformations in society and culture as lived experiences, as influenced by the very real causes and effects in the politics and economics of everyday life. They also are challenged by the realities of funded research, where most money flows from agencies that seek data and analyses to help understand what has happened and what needs to be done.

For example, in Europe today national agendas related to Europeanization and European integration have put culture squarely on the funding map. The EU has demanded that changes be made to the ways we approach culture and heritage, minority rights, and the politics of gender, ethnicity, race and other forms of identity, including regional ones. This makes ethnography into a vital tool to study and understand identity, identification and culture, as they relate to social, political and economic institutions and practices. Despite this, in many of the new EU member states anthropology is under-represented and under-funded, in part because it is not seen to be a pressing need when compared to other more longstanding disciplines. My point here, however, is not to whine about the sad state of anthropology in Europe. Anthropology in fact is growing continent-wide. Rather my intention is to continue the examination of how certain forms of scholarship face impediments that in the end hinder border and border regional research in anthropology, and, in turn, to suggest some ways that the anthropology of borders converges with the interests of other scholars through a focus on policy and history.

Borders, Anthropology and Networks

One common concern that has increasingly informed comparative border studies has been with public policy in borderlands and regions, whether it be local, regional, national or European policies. This attention to policy is with the formal institutions of power and decision-making and how they create and implement forces of regionalism and regionalization at and across borders. Comparative border studies in Europe have been based on research motored by EU and member states' interests in the forces of European integration and Europeanization. Anthropology as a discipline has a creditable record in regard to the study of policy and borders in Europe, but in comparison it has fared less well elsewhere in the world.

The anthropology of borders and frontiers that has developed over the last twenty years has often focused on issues of public policy. Most of this has been on how public policies influence if not determine much that happens at international borderlands. While this focus on policy in anthropological accounts of borderlands has at times not been particularly deep, due in large part to the growing interest by anthropologists in the issues of identity, nation and state in borderlands, it has nonetheless been a wide focus, if for no other reason than it examines the policy dimensions to border life as seemingly self-evident forces that frame or shape border experiences.

This relative dearth of policy analysis by scholars who are clearly interested in issues of politics and power is not surprising. Over the last generation of scholarship there has been a turn across the social sciences to theorizing culture and power as they relate to interpersonal relations rather than to the organs and practices of the state and its agents. These theoretical interests in power place emphasis on individuals, the body and social process. They also complement other approaches to the phenomena of globalization with the assertion that the nation is no longer what it once was and that the nation-state is in decline.

Although anthropologists have had a mixed record in regard to policy in Europe, the situation is changing rapidly and for the better. Anthropologists of borderlands have increasingly addressed policy and its relationship to state and other forms of political and economic integration. In fact, the anthropology of borders has said a great deal about policies and other aspects of institutional politics, all within various theoretical approaches (see, for example, Haller and Donnan, 2000; Heyman and Cunningham, 2004; Horstmann and Wadley, 2006; Wilson and Donnan, 1998, 2005). For example, anthropologists (such as Chalfin, 2004; Cunningham, 2004; Cunningham and Heyman, 2004) have called our attention to issues of mobility, movement and stasis at borders, and have done so with reference to national and local policies. Overall, however, these policy anthropologists, while growing in numbers, remain a minority of those who conduct ethnographic research on and theorize borders, boundaries and frontiers.

This too is changing, due in part to an inescapable aspect of field studies in borderlands. There is always a multi-locational character to border studies when one must work in no fewer than two national or regional constituencies that face each other across the political divide between states. Border studies in anthropology and in other disciplines demonstrate that the international boundary between states, for many intents and purposes, is arbitrary, and is often invisible to locals who live, work, socialize and intermarry with others across the borderline.

Thus, in one sense all borderland ethnographic work is multi-sited. But there has been so much emphasis in anthropology in recent years on the gains suggested by multi-sited research, gains linked to a speeded-up globalized world of flows and challenged boundedness, that one fact often escapes anthropologists in this rush to adopt these relatively new approaches. The majority of people with whom anthropologists still live, study, and sustain relations live in groups characterized by various forms of local intimacy and local limits. The members of these groups have known each other long before the ethnographer arrived, and they often are in regular forms of cooperation, dispute, contest, intercourse, and play, among many other behavioural practices that may give meaning and definition to the groups of which they are a part, and from which they are apart.

The people who are the hosts and interlocutors to ethnographers are immersed in webs of relations and meanings that researchers were once there to identify and understand. The people studied often know each other, or know of each other, and have done so for some time. They have in fact been involved in complicated and often sophisticated forms of society, politics, economics and culture, most of which in one way or another will continue after the ethnographer departs. These experiences and relationships are in fact why the anthropologist arrived on scene in the first place. And no matter what anthropologists assert about the lack of boundedness in culture and the fall of borders in a borderless world, most people in the world today, most of the time, and in most ways, recognize and value, whether positively or negatively, the boundaries of social and political groups.

While all ethnography since its origins has been dependent on the networks of key informants that have been established by ethnographers, the balance in evidence and argument has shifted greatly over the last generation of anthropology in favour of ethnography that relies almost entirely on the spoken words of these key respondents, with little attention paid by the anthropologist to other forms of data, data collection and data analysis. As a result of this turn away from many other forms of empirical research beyond the interview, much multi-sited research is dependent on ethnographic network building. In each site, whether it be a single location (surely with some boundaries defined by respondents and/or the researcher?) or multi-sited, anthropologists' hosts often know each other through their own networks, and invite anthropologists into the networks, or get to know each other through the network that has been created with the ethnographer at the centre, or perhaps just with the ethnographer in an important node. Their relationship to each other is often not influenced to any great extent by their contact with the ethnographer, who, as often happens, stumbles into the network, and is lucky to gain admission or access to it.

These networks of people with authority and power have long been the subject of anthropological analysis. Applied and political anthropologists have historically studied political and social elites, and utilized network analysis to chart the numbers and political influence and power of these groups. As I have suggested above, this interest in elites and their networks, in informal 'quasi-groups' and other associations, which was first and perhaps best identified by Eric Wolf (1966), is an important and welcome tactic for anthropologists who seek to temper the drive to theorize the new at the cost of using tried and true methods to identify and chronicle social and political change. While I of course accept that anthropologists have much to learn by slipping the bounds of geographically bounded space and studying transnational processes, they also have an obligation to use both old and new ways to study groups of people within geographically bounded places, in their local, regional, national and supranational contexts.

One particular difficulty to be met in the ethnography of people with power, however, is that their influence, power and high status often impede access to them as individuals and groups. They may be too busy, self-important, or distant in real and metaphorical ways for an anthropologist to get close to them, for an interview or for a more durable relationship. The most efficacious approach to political and other elites, who stay within or cross the bounds of locality, region, state and intergovernmental/supranational organizations, still seems to me to be network analysis. This method may be the most important one in the tool kit of an ethnographer, supplanting genealogical analysis as the ethnographic master method, because anthropologists seek out and use networks all the time in their field research.

Thus, in this essay I suggest that ethnographers should re-visit the concepts of networks and elites as ways to better apprehend how groups of people in the corporate, political, civic and even academic worlds gain and maintain their edge in societal organization. While many elites are localized, others are spread over a continent. All of them are part of networks that offer scholars various points of entry into the policy process, and the ways in which policy discourse, implementation, interpretation and effects frame how people and organizations are created and are interrelated. Many of them transcend borders, bringing us back to the thorny issues of how to study people who are members of groups who are not neatly situated in space and time, at and across borders.

Anthropology and Policy

The recognition that groups of people have political roles and power in society forced changes in methodology and theory in anthropology in the 1960s. One innovative way forward was the theorizing of 'social and political fields', which helped to bring together diverse approaches in applied and political anthropology at the time. These concepts were particularly helpful to ethnographers who sought to understand the formal and the informal nature of politics and power, wherever they were to be found in society and culture. Social and political institutions and organizations were made more intelligible through ethnographic attention to the many networks, informal groups, associations, and sub-cultures which provided the foundation to the sources and organizing principles of politics and policy. These institutions also framed the cultural forms through which political practices were made meaningful. And while the sub-field of political anthropology may no longer have much definition (Vincent, 1990), from the 1960s to today anthropologists have continued to investigate local level politics, both in their institutional and practical dimensions (as pioneered in the collections by Marc Swartz (1968) and Marc Swartz, Victor Turner and Arthur Tuden (1966)). Today, while so many ethnographers seek to chronicle the processes of power in their fieldwork, anthropology as a field must still accept the challenge of identifying and studying public policy, where formal politics have an impact on the lives of people in all localities.

Many anthropologists have taken up this challenge by developing new approaches in an anthropology of policy in order to at the least respond to the forces of globalization and neoliberalism that seem to pervade much public debate. However, in recognizing the importance of policy, and in particular the importance of some new policies in a rapidly changing global scene, anthropologists must remember that an anthropology of policy is not a new field of anthropological inquiry and theory (as suggested in Shore and Wright, 1997). The history of applied anthropology in the United States and elsewhere is a history of policy studies, including policy planning, implementation, effects and re-formulation. Indeed, the history of political anthropology may be usefully approached from the perspective of an anthropology of policy. James Mooney's research [1896; see also van Willigen, 1986, p.145] on the Native American Ghost Dance is a good starting point to show the intersection of applied (or in Malinowski's term 'practical anthropology'), political and policy anthropology.

In order to highlight why an anthropology of policy in Europe today might also be seen to be, at least in rough form, an anthropology that is applicable to the lives of Europeans, and also to funding and other institutions with political and economic clout, such as governments, I suggest that ethnographers must continue to consider what anthropologists *used to do* (at least in greater numbers and in more situations) more generally in their practical, applied and political anthropologies. I make this suggestion with an appended query: can or even should anthropologists do an anthropology of policy without continual if not constant reference to bounded places, spaces and groups, and political, legal and social institutions, framed within a locality, among many other real, bounded social and political units? This compounded question is particularly relevant for anthropologists and other scholars who research Europe's borders, whether from a perspective of a borderless or a bordered world.

Anthropologists will find it of great and continuing benefit to approach groups of people who cannot be easily bounded, by anthropologists or others, or who themselves seek to escape the boundaries imposed on them by others, including scholars who used to like to pigeonhole. But the benefits which I suggest will continue to accrue to those who study political fields should not be overlooked. These approaches see culture and society as the twin concepts at the core of politics, wherein social and political institutions intersect the less concrete notions of symbol and meaning, of identity and identification, and of being and belonging.

One of these old approaches which I have followed in my own research in European borderlands is that of identifying and evaluating the 'policy process' (which I borrowed from Van Willigen, 1986, p. 144). This process, which in short hand may be seen as six sub-processes, enables the scholar to show historical depth to policy studies and to demonstrate ways in which the anthropologies of policy, borders and governance intersect in today's shifting political and economic worlds. These sub-processes are integral to the determination, delivery and reception of policy, and each of them is worthy of anthropological attention in any field study of political economy. These processes are: the awareness of the need for a policy; the formulation of principal, alternative or dependent solutions; the evaluation of these solutions; the formulation of policy; the implementation of policy; and the evaluation of policy. The logic of this policy process indicates that no matter how an ethnographer chooses to identify and study policy, he or she will have to identify and understand the nuts and bolts of *policy formation*, *policy implementation*, *policy reception*, and *policy reformation*.

But these processes present certain methodological problems, among others, when the difficulties presented by geographical distances between actors are compounded by many national and other cultural differences, and by the creation of a new form of polity, a supranational European Union. New forms of governance and power are transforming polities and their policies in Europe, and are transforming the boundaries which frame them, particularly at international borders.

European Borders and Policy

Anthropologists and other ethnographers who are interested in European societies have had to deal with professional, theoretical and methodological constraints which have hindered the development of an anthropology of Europeanization and European integration. It is often difficult to utilize participant observation and other ethnographic methods in efforts to chronicle Europeanization and European integration in local communities which are distant from the centres of national and European policy-making. Moreover, an anthropology of the EU must approach in some way 'European' policy and practice in each national constituency, and how they are related to the issues of nation, state, region and locality. This is difficult to do with a methodology which in the main fixes the researcher in place and in time. That is why this essay argues for a broader sweep in spatial and temporal perspectives in the ethnography of European borderlands. This ethnographic approach contributes to the growing anthropology of European integration, which has increasingly turned to the investigation of the processes of European identification and Europeanization.

I have done research intermittently in the border region of Northern Ireland since 1991 on the effects of the completion of the EU Single European Market on cross-border economic co-operation, as well as the impact of European Commission policies on borderland political economy (Wilson, 1993, 1995, 2000). One of the conclusions I reached in this research has been that although the EU program to establish a true common market in its member states created many new cross-border realities, overall it had a relatively slight effect on local borderland attitudes towards closer political, social and cultural integration, both between the two main national communities of Northern Ireland, and between peoples in Northern Ireland and its neighbour, the Republic of Ireland. The resistance in the Northern Ireland borderlands to EU attempts to do anything beyond the strictly economic is significant in terms of the EU's wish to foster an affective dimension to its Europe-building, and may have much to tell anthropologists of the role which nationalism and strong national identities may play within the wider processes of Europeanization in Western Europe, in a EU which recently expanded to twenty-seven members. This tension between the affective and instrumental dimensions of Europeanization as they are experienced in Europe's localities may also shed light on how Euroscepticism has developed in the new member states of Central and Eastern Europe (Ilieva and Wilson, 2011).

Overall, the EU has achieved little in the Northern Ireland borderlands in terms of adapting to the EU in ways more than economic, or to transnational cultural integration allied to parallel processes of deterritorialization, globalization and regionalization. In the border region European identity is not a widespread or active social identification, among people who are mostly members of the nationalist, Irish and Catholic traditions in Northern Ireland, and is seldom a focus of everyday discourse. Locally, Europe takes a back seat in most arenas of public life to the social, cultural, economic and political agendas that in the past were about supporting or subverting the local economy and the nation and state, and today are more about maintaining the peace which has been in place in Northern Ireland for fifteen years. In my research in many locales in Ireland this relative absence of identification with Europe has even been growing among farmers, who have long been held to be among the most pro-European groups in both the Republic of Ireland and in Northern Ireland.

While it may not be surprising that in the conditions of nationalism that have existed in Northern Ireland for generations that the instrumental dimensions of European integration trump those of the affective, some scholars have looked precisely at borderlands in Europe as the arena within which new forms of European identity will take shape. Kohli (2000), for example, has suggested that border peoples will be among the first and best supporters of European identity within the processes of European integration, due to their years of mixing national and other identities at the frontiers of their nations and states. Such calls aside, in the Northern Ireland borderlands, despite many years of development policies aimed at benefiting borderlanders, these policies have had a negligible impact in creating a stronger affective identification of local people with the Europe of the EU.

Policies and other forms of institutional involvement in Ireland have had an impact on aspects of local life in the borderlands that go beyond issues of identity and can be monitored using the policy process privileged above. European integration has facilitated the creation of cross-border funding networks. Because so much European funding in both Northern Ireland and in the Republic has been about the establishment of partnership, across sectarian and national lines and across the international borderline, Europeanization has resulted in new policy networks and elites. District Councils in Northern Ireland and County Councils in the Republic have created new forms of governmental cross-border cooperation, complementing the North South Ministerial Councils that were created in the Belfast Good Friday Agreement. These councils and committees often meet in the borderlands. While much of this cooperation is in policy areas, such as tourism, which are unlikely to inflame nationalist or unionist emotions, the mere fact of their existence has opened up new policy dimensions for the borderlands.

Professional consultants are another elite network that has filled a niche created through EU policies in the borderlands. These economic and political advisors offer their expertise to people in the public and private sectors who seek European funding. These consultants, many of whom have worked in local and national government or administration, are seen by local businessmen and community groups as preferable alternatives to civil servants, even though this latter group offer the same service at no cost and as a right to citizens. The expressed preference to pay for such services is largely the result of two factors: the distrust of sharing economic data with the government in any circumstance, as it may find its way to the Inland Revenue, and the fact that local government agencies are often competing for the same European funds that they advise their constituents to seek! Private legal, accounting and political affairs firms provide confidentiality, as well as the promise of insider knowledge, due to the networks former government employees created when they were in public service.

These two examples are offered as an indication that in these Irish borderlands, the study of policy formation, implementation and reception must minimally consider the continuing roles of economic and political networks, some of which may be old or new elites, who in various ways affect policy. Moreover, my research in the Irish borderlands also provides evidence of how the affective and instrumental dimensions to Europeanization should be considered in tandem with the study of European and national policy. In the conclusion I offer some thoughts as to how ethnography might best serve the overlapping goals of border, policy and political studies in anthropology, as well as how anthropology may continue to contribute to wider comparative border studies.

Conclusion

In the early 1990s, when I was working at the European Studies institute in Queens University in Belfast, my friend in the social anthropology department and I decided to offer 'the comparative study of borders' as a conference theme to the Anthropological Association of Ireland. We organized that conference, which had a particular emphasis on the land border between the Republic of Ireland and the United Kingdom, in Northern Ireland. The conference was a success, largely because of the venue in the border village of Carlingford in County Louth, but as we were working on the book that we had agreed to edit of the conference proceedings, we laughed to ourselves as we wondered if there would be any interest in our book, and we played around with the notion that if we had arrived at borders as a research topic and arena for comparative anthropological investigation, then perhaps all that had been interesting on the subject had already been said! Well, that book (Donnan and Wilson, 1994) was the first of six books we have written and edited on borders, and we are currently editing *A Companion to Border Studies* (Wilson and Donnan, 2012). We were wrong to conclude that much that needed to be said had already been said in 1994, if the breadth and depth in border studies over the last twenty years is any evidence. Borders may be as fascinating today to scholar, politician, journalist and person on the street as ever they were.

The popularity of the subject might be one factor that keeps border scholars on the subject of borders, but it is only one, even when it is harnessed to scholarly interest in such relatively compelling subjects as citizenship, sovereignty, nationalism, imperialism and migration. There are other factors that may very well connect border studies scholars to each other, and in this essay I have reviewed some of those most important to social anthropologists, or at least to those who work on the overlapping issues of state, nation, sovereignty, territory and history. In Europe if not also elsewhere these issues lead scholars to the wider concerns of how borders function within the borderless world which is often asserted to be a defining feature of globalization despite the importance of borders to all forms of inter-, trans-, supra- and nationalism.

Policy also continues to be an important entry point into any study of local to global change at borders. However, in a world with some new and many shifting parameters of power and order, and with many continuing structures of governance in and beyond the state, an anthropology of policy is much more than a good entry and exit point in border studies. Public policy is itself constitutive of so much that is at the core of what borders do and mean in contemporary Europe, and its scholarly study offers all social scientists the opportunity to see borders in wide comparative perspective. But an anthropology of policy must be firmly based on ethnographic research, and ethnography must be rooted in historical reflexivity, in a manner similar to that called for by O'Dowd in regard to national borders within a re-bordered world. An anthropology of policy, which from my vantage point needs to utilize methods to identify networks and elites, is as dependent on historical reflexivity as any other study of geopolitical life, in terms of the dialectics between and among ideal and real, local and global, and borderless and bordered worlds.

Ethnography today, however, has shied away from historical reflexivity. As a result it has become diluted as a means of doing policy and political research in borderlands, and has diminished its capacity to converge with other social sciences in comparative border studies. The more that ethnographers have become the central feature of ethnographic research, the more 'anthropologists have also all too easily lost sight of the *partiality* of their participation in the world they study. They begin to believe they *are* the world they study or that the world revolves around them.' (Burawoy, 2003, p. 673)

One way to escape from this self-made and professionally sanctioned anthropological trap is for ethnographers to revisit their research. As the sociologist-ethnographer Michael Burawoy, who has sought to strike a balance in ethnography between the forces of realism and constructivism, has suggested, 'An ethnographic revisit occurs when an ethnographer undertakes participant observation, that is, studying others in their space and time, with a view to comparing his or her site with the same one studied at an earlier point in time, whether by him or herself or by someone else' (Burawoy, 2003, p. 646). In this revisiting, there must be a historical dimension, to the ethnographic data collected and analysed by the ethnographer to be sure, but also to the forces external to the researcher that are dynamic, fluid and unpredictable. Simply put, life moves on, things change if not fall apart, and ethnographic research over time, conducted by the same ethnographer or others, in the same or parallel field sites, provides the veritable stuff of historical and ethnographic reflexivity that are the lifeblood of critical social science.

In this essay I have ranged over three topics that I have related to border studies in Europe today: the anthropology of borders, the anthropology of policy and historical and ethnographic reflexivity. If border studies scholars seek in any shape or form to identify and understand culture, identity and power in borderlands, then ethnographic research provides unequalled methodological possibilities. But ethnography today is fraught with idealist notions of culture and community on a par equal to those of nation and state that are at the heart of so much post-modernist, neoliberal, and globalization rhetoric. If ethnography is to continue to serve as a methodology that draws border studies scholars together, because of common and shared interests in the cultures of mobility and stasis at and across borders, then it must also persist in its adoption of methods that allowed ethnographers to understand politics and power, such as those regarding networks and elites. But ethnography also needs to be historically reflexive and recalibrate its own balance between space and history.

References

Balibar, E. (2009), 'Europe as Borderland', *Environment and Planning D: Society and Space* 27: 190–215.

Burawoy, M. (2003), 'Revisits: An Outline of a Theory of Reflexive Ethnography', *American Sociological Review* 68(5): 645–79.

Chalfin, B. (2004), 'Border Scans: Sovereignty, Surveillance and the Customs Service in Ghana', *Identities: Global Studies in Culture and Power* 11(3): 397–416.

Cole, J.W. (1977), 'Anthropology Comes Part-Way Home: Community Studies in Europe', *Annual Review of Anthropology* 6: 349–78.

Cole, J.W. and Wolf, E.R. (1974), *The Hidden Frontier: Ecology and Ethnicity in an Alpine Valley*, New York: Academic Press.

Cunningham, H. (2004), 'Nations Rebound? Crossing Borders in a Gated Globe', *Identities: Global Studies in Culture and Power* 11(3): 329–50.

Cunningham, H. and Heyman, J.M. (2004), 'Introduction: Mobilities and Enclosures at Borders', *Identities: Global Studies in Culture and Power* 11(3): 289–302.

Donnan, H. and Haller, D. (2000), 'Liminal No More: The Relevance of Borderland Studies', *EthnologiaEuropea* 30(2): 7–22.

Donnan, H. and Wilson, T.M. (eds) (1994), *Border Approaches: Anthropological Perspectives on Frontiers*, Lanham, MD: University Press of America.

Donnan, H. and Wilson T.M. (1999), *Borders: Frontiers of Identity, Nation and State*, Oxford: Berg.

Haller, D. and Hastings, D. (eds) (2000), *Borders and Borderlands: An Anthropological Perspective*, Special issue of *EthnologiaEuropea* 3 (2), p. 126.

Hansen, E.C. (1977), *Rural Catalonia under the Franco Regime: the Fate of Regional Culture since the Spanish Civil War*, Cambridge: Cambridge University Press.

Heyman, J.M. (1994), 'The Mexico–United States Border in Anthropology: A Critique and Reformulation',*Journal of Political Ecology* 1: 43–65.Available at: http://www.library.arizona.edu/ej/jpe/volume_1/HEYMAN.PDF [accessed 7.7.2009].

Heyman, J.M. and Cunningham, H. (eds) (2004), *Movement on the Margins: Mobilities and Enclosures at Borders*, Special issue of *Identities: Global Studies in Culture and Power* 11(3).

Horstmann, A. and Wadley R.L. (eds) (2006), *Centering the Margin: Agency and Narrative in Southeast Asian Borderlands*. Oxford: Berghahn.

Ilieva, P. and Wilson, T.M. (2011), 'Euroscepticism and Europeanisation at a Margin of Europe', *Anthropological Journal of European Cultures* 20(2): 87-113.

Kohli, M. (2000), 'The Battlegrounds of European Identity', *European Societies* 2(2): 113–37.

Mann, M. (2007), 'The age of nation-states is just beginning', Unpublished paper, *Beyond the Nation* conference, Queens University, Belfast, 12 September.

Marcus, G. (2007), 'The Passion of Anthropology in the U.S., circa 2007', *Anthropological Journal of European Cultures* 16(1): 29–55.

Mooney, J. (1896), *The Ghost Dance Religion and the Sioux Outbreak of 1890*, Washington, DC: Bureau of American Ethnology.

O'Dowd, L. (2010), 'From a "Borderless World" to a "World of Borders": Bringing History Back in', *Environment and Planning D: Society and Space* 28: 1031–50.

Rumford, C. (2008), *Cosmopolitan Spaces: Europe, Globalization, Theory*, London: Routledge.

Schneider, J.C., and Schneider P.T. (1976), *Culture and Political Economy in Western Sicily*,New York: Academic Press.

Schneider, J., Schneider, P. and Hansen, E. (1972), 'Modernization and Development: The Role of Regional Elites and Noncorporate Groups in the European Mediterranean', *Comparative Studies in Society and History* 14(3): 328–50.

Shore, C. and Wright, S. (1997), 'Policy: A New Field of Anthropology', in C. Shore and S. Wright (eds) *Anthropology of Policy*, Routledge: London.

Swartz, M. (ed.) (1968), *Local Level Politics*, Chicago: Aldine.

Swartz, M., Turner, V. and Tuden, A. (eds) (1966), *Political Anthropology*, Chicago: Aldine.

Van Houtum, H. (2005) 'The Geopolitics of Borders and Boundaries', *Geopolitics* 10: 672–9.

Van Willigen, J. (1986), *Applied Anthropology: An Introduction*, South Hadley, MA: Bergin and Garvey.

Vincent, J. (1990), *Anthropology and Politics*, Tucson: University of Arizona Press.

Wilson, T.M. (1993), 'Frontiers Go but Boundaries Remain: The Irish Border as a Cultural Divide', in T.M. Wilson and M.E. Smith (eds), *Cultural Change and the New Europe: Perspectives on the European Community*, Boulder, CO: Westview Press.

Wilson, T.M. (1995), 'Blurred Borders: Local and Global Consumer Culture in Northern Ireland', inJ.A. Costa and G.J. Bamossy (eds), *Marketing in a Multicultural World: Ethnicity, Nationalism and Cultural Identity*, London: Sage.

Wilson, T.M. (2000), 'The Obstacles to European Union Regional Policy in the Northern Ireland Borderlands', *Human Organization* 59(1): 1–10.

Wilson, T.M. and Donnan, H. (eds) (1998), *Border Identities: Nation and State at International Frontiers*, Cambridge: Cambridge University Press.

Wilson, T.M. and Donnan, H. (eds) (2012), *A Companion to Border Studies*, Oxford: Blackwell.

Wilson, T.M. and Donnan, H. (eds) (2005), *Culture and Power at the Edges of the State: National Support and Subversion in European Border Regions*, Münster: LIT Verlag.

Wolf, E. (1966), 'Kinship, Friendship and Patron–client Relations in Complex Societies', in M. Banton (ed.) *The Social Anthropology of Complex Societies*, London: Tavistock.

Wolf, E.R. (1982), *Europe and the People without History*, Berkeley, CA: University of California Press.

Chapter 5

Borders from the Perspective of 'Ants': Petty Smugglers from the Polish–Ukrainian Border Crossing in Medyka

Marta Byrska-Szklarczyk

Introduction

In this article I am going to present the Polish–Ukrainian border from the perspective of the community, whose life is organized around its everyday experiencing of the border. The protagonists of my research are so-called 'Ants'[1]: petty cigarette smugglers operating at the border crossing point in Medyka, who earn their living by carrying cheap, often forged Ukrainian cigarettes through the border and selling them with much profit on the Polish side. They are inhabitants of the surroundings of Przemyśl (a town about 12 kilometres away from the border), for whom the opening of the borders in 1989 and the subsequent economic changes, which led to the collapse of many factories in the region, meant the border crossing is the only 'workplace' left.[2] In 1998 the pedestrian crossing was opened in Medyka – the road and railway ones already functioned – and after Poland joined the European Union in 2004 it has become the biggest crossing point of that kind separating the EU and 'the rest of the world'. At the time of my field research the number of Ants at the pedestrian border crossing fluctuated between 6000 and 12000 per day (Delmanowicz, 2008a, 2008b). Other users of the crossing are rarely encountered. The very term 'Ants' descriptively refers to a great number of petty smugglers crowding at the border day and night as well as to their mobility, small noxiousness

1 The term functions in the Polish language both in a colloquial speech and in the discourse of the media and politics but it is also a means of internal identification.

2 The phenomenon of petty border trade and ant-like cigarettes smuggling (also called 'trader tourism' or 'shuttle trade') is characteristic to the Eastern regions of Poland. Differences in prizes of cigarettes, alcohol and fuel between Poland and the neighbouring countries are the reason why such practices can be encountered at all border crossings (cf. Matejko, 2008; Gołębiowski et al., 2009; Cegliński, 2005; Gołębiewski, 2010; Jagnyziak, 2010; Plińska, 2010; Chlebika, 2010; Kazimierska, 2006). These practices inscribe into the broader, anthropologically characterized phenomenon called 'Cross-border small-scale trading' or 'Trader-Tourism' (cf. Konstantinov, 1996, 1998; Hann and Hann, 1992; Henrik, 2006).

of their ant-like practices (in comparison with the scale of 'truck smuggling') and their ability of effective cooperation aiming at the fastest possible crossing of the border and overcoming obstacles created by the state administration.

Ants earn money due to the legal loophole – until December 2008 Polish customs law allowed one to bring into the territory of the state one carton of cigarettes for one's own use. The people who 'carry the norm' – as those crossing the border with one 'legal' carton are colloquially referred to – in fact do not violate the law, but earn by crossing the border with the goods a few times a day (the most cunning managed to cross the border even up to five times a day). Many Ants were not satisfied with the 'norm', therefore they decided to risk smuggling additional cartons, packs or even single cigarettes. These were obviously not purchased for their own use as Ants declared to the customs officers but with the intention to sell them on the Polish side to so-called 'Buyers' ('Skupczycy'), who are the first link of the mob-like organization which smuggles forged Ukrainian cigarettes to the West, mainly to Germany, where, first re-packed into German boxes, cigarettes are sold in street vending machines. Representatives of the local administration: customs officers, customs soldiers and politicians were perfectly aware of the shady business. However, due to the mere social noxiousness of this type of 'trade', poverty and enormous unemployment in the region, Ants' activities were silently approved by the authorities.

The border itself is of fundamental importance to the existence of the local community whose everyday experience adds to the creation of a specific 'culture': local language, rituals, perception of space, time and physicality, power relations and economic transactions. I will try to depict the complexity of various contexts, imageries and mythologies contained in the local knowledge of the border through an analysis of the metaphors used by the Ants which I will treat not as a means of stylistic expression or a tool of anthropological description, but as a bridge between my own world and the world of the protagonists of my research (Hastrup, 2008 [1995]).

The aim of this article is therefore to present a great variety of local metaphors of the border: images, statements and experiences contributing to what the border is for people who experience it in their own body and language on a daily basis. Later in this text I am going to elaborate on the most popular metaphors of the border: metaphors of life and death of the border, the border understood as movement, the border as a circle and a factory, the border as a meeting point and the end of the world and the border being a prison and a war front.

Through the presentation of the subsequent metaphors I intend to describe everyday life of Ants in two periods which were breakthroughs for the local community. The first of them covers the time after Poland joined the Schengen Area in December 2007 when Ukrainian Ants disappeared from the border crossing as a consequence of the expensive and complicated procedure of granting new 'Schengen visas'. For the Polish Ants the crossing still functioned the 'normal' way. The second period refers to the events which took place in December 2008 after the introduction of the new customs regulations which limited the number of cigarettes that can be legally 'brought' through the border to two packets, instead of a previously allowed carton which consequently made Polish Ants' border activity unprofitable. In response to the new regulations Ants organized a demonstration which blocked the crossing point, then spent several days and nights protesting. I took part in the above events during the second research I was conducting at that time.

Methodology

My fieldwork at the border was conducted in two stages: one of the travels took place in May, the other in December 2008[3]. In both cases I collected materials by means of ethnographic interview and participating observation which are classical methods of anthropological research (Clifford, 2000, p. 42; Hastrup, 2008 [1995], pp. 60–61). As Kirsten Hastrup writes:

> Although we obviously cannot experience the world from other people's perspective, we can share their social experience [...] Sharing then means that we are part of the events which are fundamental from the point of view of the local community. (Hastrup, 2008 [1995], p. 65, translation by the author)

This idea of sharing significant (but also trivial) events which my interlocutors are involved in, being with them, accompanying them in their everyday activities, talking with them, listening, watching, experiencing, eating and smelling what they do, had always been a methodological credo for me.

Before I began my research at the border crossing point in Medyka, I had frequently visited this place as a tourist travelling to Ukraine and I had crossed the border in different ways: walking, on a bicycle, by bus. I observed hundreds of people – mostly older women – crowding at the border at night in order to gain less than twenty zlotys selling cheap cigarettes, and I had personally experienced physical pain and humiliation in the sinkhole (a place of long-time waiting for the customs check-up) which contributed to the choice of my current anthropological explorations.

In May 2008 when I was doing my first field research, I was not able to be much involved in smuggling practices of Ants due to pregnancy. My condition which at first seemed an obstacle on the way to conduct ethnographic research – I couldn't share the most important experience which was smuggling itself – eventually turned out to be extremely useful in breaking cultural barriers. The initial distance resulting from the fact that I was treated as a 'privileged' young student coming from the

3 The research has been financed by the Ministry of Science and Higher Education within the research funding *Ukraine: Memory, identity, subjectivity* (1 H01H 011). As a result of the project, the book *On the border of the new Europe. Polish-Ukrainian neighbourhood*, ed. Magdalena Zowczak was published (Zowczak, 2010). The book contained an article which is further developed and extended in this text (Byrska-Szklarczyk, 2010).

capital who was interested in 'them' functioning half-legally and as economically excluded citizens of the poorest regions in Poland, was overcome thanks to a common and uniting experience of feminine corporeality and my own physical suffering which I often could not hide. The great majority of Ants were middle-aged women most of whom had experienced motherhood. The unity of identically alive, 'procreating' and suffering bodies as well as traditional respect towards motherhood made me closer to those women, more human and less 'from Warsaw'. My visible pregnancy 'made it' even further. Namely, from the first day of my research Ants encouraged me to practice so-called 'pregnancy smuggling'. It is one of the popular strategies of crossing the border quickly, without having to wait in a long queue or being checked thoroughly by a customs officer. Pregnant women from Przemyśl and the surrounding villages eagerly use such privileges, similarly to people in wheelchairs, with crutches, and those who are blind or mentally handicapped. People who are stereotypically perceived as suffering, harmed or excluded, have a high status here: they are 'fully capable' Ants because at the border you are 'fully capable' if you can smuggle fast and a lot and cunningly cheat customs officers. Being handicapped or pregnant becomes a game, 'an official version of events' for customs officers. Physicality and 'disability' is creatively used and reinterpreted. Besides, these roles are largely conventional. I frequently heard an opinion that it's hard to tell who really is 'abnormal' and who only pretends mental or physical disability in order to be more effective on the border. How does that relate to my deliberation on the condition of an anthropologist 'observer and participant'? The most crucial element of the situation described above is the fact that from the very first day I felt being admitted to the local practices and notions. My pregnancy was read in the local system of meanings which places 'being pregnant' high in the hierarchy of the smuggling potential. Despite my initial fears I managed to 'be in the field' to share and experience local practices and 'important events', as described by Hastrup (2008 [1995]).

The second field research which took place in December 2008 was of quite a different nature. Having learnt about the change of the customs regulations resulting in Ants' protests on the border, I spontaneously decided to join the demonstrations. This time it was much more difficult for me to win their confidence. Chanting slogans against strike-breakers, politicians, customs officers, administration workers and the police, Ants initially perceived me as 'sent by the enemy', a 'secret agent', a 'fake customs officer' or a 'tabloid journalist' who wanted to use their misery to prepare material for the newspaper. Before I finally gained their trust, I had to show my documents and various certificates from the university several times and persuade Ants ad nauseam that I only wanted to hear their opinions on what it means to live at the border. I managed to gain their trust only when I talked to the local leaders, so-called 'Old Ants' who were women most actively involved – once in smuggling – now in the protests against sealing of the border. Finally, I could stand with Ants on the border days and nights, often in the cool December rain, listening to their stories, wailing and insults thrown at the whole world, drinking tea from one bucket, eating the cheapest sausage to be found in the local supermarket and singing songs by the fire arranged in steel barrels. I also participated in the meeting with representatives of local politicians which was meant to find a solution for the disastrous situation of Ants; after changing of the customs laws by the government suddenly a few thousand people had been deprived of their only source of income. At this meeting I had an occasion to experience a clash of two completely different discourses of the border using two completely different ways of expressing and imagining it.

In this article I will be using the term 'discourse' in reference to Paul Ricoeur's concept of discourse understood as the dialectic of event and meaning (Ricoeur, 1976, p. 12). Ricoeur argues that: 'If all discourse is actualized as an event, all discourse is understood as meaning'. Analysing the local discourse of the border I will attempt to describe the ways of its 'happening' (event) and to decipher the meanings it carries. (Ricoeur, 1989, pp. 79–85)

Let me elaborate on the key concept to this work: the concept of the metaphor. By talking with Ants and overhearing their conversations I experienced momentary (limited to my visit) immersion in the local discourse of the border. I noticed that Ants present their everyday life at the border not straightforwardly but in a metaphorical way by building bridges between different categories of reality and language. In their descriptions of the border they refer to other areas and meaning categories in order to create comprehensible – also to me – areas of references. By comparing the border to a prison or a concentration camp they place themselves in an intelligible and more universal system of meanings. For an anthropologist who, despite a long-term presence in the field, is never going to become 'one of the group' he or she examines, common points of reference become crucial. In such cases metaphor becomes an extremely useful tool of ethnographic transposition which makes it possible for an anthropologist to evoke a local vision of the world, to understand it and present it anew by means of the academic discourse (cf. Hastrup, 2008 [1995], p. 33).

In this article I would like to make twofold use of the metaphor potential: as a method to place the argument in a logical order which is a tool of anthropological textualization and to give voice to Ants, in an attempt to fulfil the idea of dialogism or polyphony in ethnography (Clifford, 2000, pp. 53–63).

Metaphor is much more than just 'wordplay' or a means of stylistic expression. As Hastrup writes: 'metaphors are not about comparing … two semantic fields and defining similarities between them; metaphor has only one subject' (Hastrup, 2008 [1995], p. 50, translation by the author). For Ants, the border does not 'resemble' a prison or a bazaar but is actually experienced as a prison, a bazaar, a front etc. Metaphor is more than a means of shaping local discourse of the border. It is a way of thinking, acting, understanding and experiencing reality.

The Life and Death of the Border

The key metaphor covering the whole period of my research and an important stage of Ants' activities in Medyka is a metaphor of the life and death of the border. With the help of this metaphor Ants told me the story of their border mainly during the protest, i.e. from the perspective of its 'end'. What politicians and the media called 'sealing of the border', was locally considered as its death. The period of time that ended in December 2008 was regretfully recalled as a period of wonderful prosperity: 'This was normal selling, everything was alive.' (Female unemployed Ant aged 40). The border itself was presented as a lively meeting point, a local centre. Ants thought of themselves as the propeller not only of this place but also of the whole economically collapsing region. In their narratives a symbolic pack of cigarettes carried through the border by an Ant and put into further circulation becomes a guarantee of the local world's existence. Ants drew in front of me a coherent concept of the region as if it functioned solely thanks to their work. They make trade in the border bazaars flourish (in Ants' statements 'trade' frequently becomes a synonym of life, happening, movement), the exchange offices work (three currencies function at the border: zlotys, dollars and hryvnias) and buses transporting people from remote villages to Medyka make their everyday routes. In order to understand the sense of the metaphor of life and death of the border, I am going to present more in detail the specificity of 'life at the border'.

Medyka is divided into two parts: Medyka the borough and Medyka the border crossing. Around 2000 citizens live in the village itself. Most of them live in country houses and post-State Agricultural Farm blocks of flats, some of them own potato and hop fields. There are two churches and a dilapidated synagogue from before World War II which serves as a car park for the border fire brigade. On the road both an old cart and Volkswagen Passat can be observed which is the most popular car model in the border areas due to the very capacious fuel tank. The majority of citizens of Medyka come to the border each day either as Ants or their clients-retailers (so-called Handsellers). Others visit the border for entertainment: to talk, have a beer, and watch the border life. The railway tracks symbolically divide Medyka into the village and the border area.

The biggest bazaar in the border area is 'Centrum Handlowe Granica' ('The Border Trade Centre') where it is possible to buy almost anything: from the Ukrainian products on which excise tax is levied such as cigarettes and alcohol, through food, underwear and household equipment to various medicines of most suspicious origin such as bear fat (it's worth noticing that the animal is a protected species). Most of stands and stalls at the border offer products that are expensive or unavailable on the Ukrainian side. While the Polish buy alcohol, cigarettes and sweets in Szeginie', the Ukrainians purchase oil, onions, chicken feet, coffee, sheets, pots and sausages in Medyka. Apart from the bazaar, in the area around the border crossing point there is a huge car park, a cheap supermarket coated with siding or panels, extremely smoky and dangerous pubs where one can lose all the money previously earned at the border on slot-machines. These bars are only (as some mentioned to me in confidence) a mere cover for the illegal business of cigarettes smuggling. Supposedly, illegal goods are stored in huge amounts in cellars and walls of those bars waiting to be smuggled to Poland and further to the West. The life in Medyka seems to be focusing around those places. On the surrounding benches Ants take rest before going to Ukraine, Buyers ('Skupczycy') are occupied with their illegal business of buying cigarettes, Handsellers sell cigars ('cygary'), there is music coming from the bars and there is always something interesting going on. The so-called 'under the border' area seems to be quite disordered as there is garbage everywhere, mainly empty cigarette packs and boxes. Pieces of adhesive tape, used by Ants for packing, spin in the air and the ground is covered with cigarette-ends, bottles and plastic beer glasses. Dogs from the village come here to feast on food remains. From the windows of the bars one can see three flags, the Polish, Ukrainian and the EU flag, stuck on the Medyka-Szeginie border buildings. 'This is the end of the EU' – I often heard Ants say.

In front of the bars, buses stop. They bring people who immediately proceed in the direction of the pedestrian crossing. Most of them carry black plastic bags with a golden inscription 'BOSS' written on them (referring to the popular clothing brand) which are broadly called 'smuggler-bags' and are used to carry through the border the legal 'norm' that is one carton of cigarettes. Going to Ukraine – 'empty-handed' – takes much less time than the way back. In Szeginie, in one of many shops offering alcohol and cigarettes (paid in US$) Ants can purchase the 'norm'. One carton (i.e., 10 packs) of the cheapest Prim cigarettes (also called rubbish) or Classics cost around $2. Ants not only buy a carton but they also get a bottle of vodka (profits are minimal but since it's within the 'norm', it would be 'a sin to make a dummy run'). If they decide to risk smuggling additional packs or cartons, they need to choose the right strategy. The most popular smuggling methods are 'klin' ('in wedge') and 'plastry' ('plaster'). The first one means to wedge several packets (or even a few cartons as record-holders do), in the fly of the trousers or in one's underwear. Cigarettes then fit close to the fragments of the body which the customs officer is not allowed to touch during the routine check-up. Obviously, if an Ant is suspected to be a smuggler, an officer can order a personal check (so-called 'rozbieranka'), but considering very long queues and the time pressure under which customs officers work, this type of control is frequently avoided. 'Plaster' strategy consists in pasting the cartons of cigarettes to the naked body with a thick adhesive tape – plaster. With this strategy it is advisable to put on loose clothes, a few layers if possible (this method is most effective in winter), so that a customs officer would not be able to feel the goods during the routine check-up. In Szeginie nobody is shocked by a view of half-naked elderly women pasting cigarettes to their bodies with each other's help or that of the local sellers. Other popular smuggling strategies consist in carrying goods in a cap or in shoes (single packs of cigarettes are sold by the piece to inhabitants of Medyka). As already mentioned in the methodological part, Ants also use 'pregnancy' or 'cripple' techniques. The latter means not only a privilege of avoiding the queue but also makes it possible to smuggle cartons underneath the wheelchair or single cigarettes in crutches. Having purchased the goods an Ant undergoes the Ukrainian check-up finalized by handing an officer a one-dollar bribe for each carton, then awaits the Polish check-up for hours in the so-called 'sinkhole', a narrow corridor surrounded with steel rails, where the crowd awaits the Polish customs check-up''. After the successful passing of the border – referred to as 'groping', 'combing', 'striking' in search of 'fags' – the Ant sells the goods to a Buyer in Medyka.

The community of Ants is not homogenous. They are the inhabitants of the countryside and of the cities, women and men, the young and the old, unemployed for whom the border is the only source of money as well as employed (both legally and illegally) who only earn extra money to heighten their low income. Among Ants I met a labourer, a farmer, a hairdresser, an administration officer, a car park employee, a student, a scholar, a barmaid, a cleaner, a pensioner. At the weekends the 'school', i.e. schoolchildren and students appear at the border. In Medyka there are three coexisting groups involved in the cigarette trade: Ants, Buyers and Handsellers. The biggest community of several thousand Ants differ internally. Its core is formed by so-called Old Ants, mostly unemployed middle-aged women who have been working at the border full-time and for a long time. They are acquainted with everybody and they know everything. They were also the most active participants of the December protests. While being an Old Ant somehow constitutes their identity (also the professional one), for many 'temporary' or 'visiting' Ants work at the border is only a method of earning additional money. In this hierarchy so-called Dodgers occupy the lowest position. These are groups of young boys who do not obey the rules worked out by Ants regarding the behaviour in the sinkhole. They tend to push through people waiting in queues, use audacious smuggling methods such as running through the customs office without undergoing a check-up or being violent towards customs officers. Old Ants have thoughts of 'eliminating them from the border' (male car park employee and Ant aged 60) because, even though they are in minority, they negatively influence the perception of Ants in Poland. There is polarization between the ones for whom the border signifies hard work and the only source of income and those for whom the border is a place of entertainment, showing off and teenage competition.

Despite the significant diversity Ants seem to be a community cemented by their identification with the region, homogeneity of actions, self-sufficiency and the awareness of being different (Rapport and Overing, 2000, p. 60). The most important features which make them a community are participation in the same linguistic reality and common experiencing of the border which shape a specific perception of the world (cf. Cohen, 1985).

While the border crossing and the sinkhole are dominated by Ants, the described above space in Medyka with bars and exchange offices is dominated by Buyers. They are young, aggressive and often drunk men who hold real power over the place. They bully Ants coming back from Ukraine and by shouting and swearing loudly force them to sell cigarettes at the lowest possible price. They do not respect Ants regardless of their age and sex. They evoke fear due to their alleged connections with the mob. The area around the bars is also occupied by

Handsellers. They are Ukrainian citizens of Polish origin who – unlike Ukrainians – do not need a Schengen visa to cross the border. They are mostly older women who hold the offered products – a carton of cheap cigarettes, a bottle of vodka, a few beers in plastic bottles, well-liked Ukrainian candies – in hand (hence the name). Their clients are usually inhabitants of the neighbour villages who frequently order specific goods from 'their' regular sellers. Handsellers also do their business in Przemyśl where they have their permanent selling places in the vicinity of train stations, city bazaars or squares.

During my first field research which took place after Poland had joined the Schengen Area, the period of flourishing and intense life of the border was gradually coming to an end. I could more and more often hear that 'the border is dying' and that 'Schengen kills'. The reason of this slow agony of the border was the exodus of both Ukrainian Ants and Ukrainian clients. The border bazaars became deserted, supermarkets in which the Ukrainians used to shop, started to dismiss employees and one would less and less often hear the lilting Ukrainian language in the bars and shops by the border. For the Polish Ants the border eventually 'ended' one year later, in December 2008. The metaphor of the death of the border was expressed by the protesting Ants not only verbally, but also through specific funereal practices. The terminal gates, where a permanent group of protesters sang sad songs, were hung with black ribbons and dried flowers 'due to mourning' as one of the Ants explained. Grief-stricken Ants expected the worst for the inhabitants of the region: 'prisons will get filled and people will commit suicides' (Female Ant aged 40). In their metaphorical statements and rituals the border emerges as a being which undergoes the same rhythm of life and death as people who live around it. The border pervades into their bodies which are doomed to decay without it.

'The Border is Made for Walking' – The Border as a Movement, a Circle and a Factory

The metaphor of life and death relates to another metaphor of the border crucial for the local discourse, which is the metaphor of a movement and a standstill. The former I only understood during the Ants' demonstrations. The protest itself consisted in a long-time motionless standing of a group of people in front of the gate to the pedestrian crossing terminals. This way the Ants manifested their refusal to 'walk across the border' and those who broke the strike rules and the Ants' solidarity by 'going to Ukraine' were called offensive names and condemned as strike-breakers.

The Ants did not block the way, they simply stood in the same place where earlier they had been in the constant movement. The media only showed a few clashes with the police while the demonstrative 'unwalking' of the Ants was ignored as senseless and illogical. The meaning of this immobility is comprehensible only in the context of the border described by the Ants as the biggest workplace in

Podkarpacie. The Ants, who are prevailingly the unemployed dismissed from bankrupted factories of Przemyśl and its surroundings, referred to the border in terms of a well-functioning factory which supports entire families. The Ants described their work as 'walking'. They said, for example: 'I have been walking here for two years', 'I do not want to, but I have to walk to earn my bread', 'Everybody is walking here' or even, 'The border is made for walking'.

The border-factory is in the constant movement pictured as a movement in a circle. This local image of the border shows how inadequate to the description of reality is the common image of the border as a line on a map. Here one does not say 'to cross the border', but 'to walk the border' and 'turn around the border'. The route from Medyka to Szeginie and back is referred to as 'making a circle'. For Ants the border is not a line separating two countries, but a circle connecting the two most important points of the economic route: the entrance to Ukraine, the Ukrainian shops, the Ukrainian customs office, the sinkhole, the Polish customs office, the bazaar in Medyka, the entrance to Ukraine – which repeats to infinity. It is not – as in the common imagery – the end or the limit of the country, but the centre, the middle of the world, the 'axis mundi'. The following remarks might be examples of such thinking: 'Medyka is known in the entire world', 'There is only one pedestrian [border crossing] – ours'.

Next to 'walking', 'carrying' or 'bringing' are other local terms for the work of the Ants. During my first visit to the border I was astonished at the way Ants talk about their transactions. One does not mention buying and selling Ukrainian cigarettes to Buyers (mob people) – the word 'smuggling' is particularly avoided – but 'bringing' or 'handing over the carton'. The goods are not called literally: cigarettes but with tenderness: 'a [little] carton' or 'a [little] package' (more seldom 'a block' or 'a bar'). The entire transaction seems to be barter – an exchange made without using money. The basic medium of exchange is a carton ('kartonik') which can be 'turned into' various goods.

A carton was exchanged 'for bread', 'for clothes', 'for children's' studies', 'for life'. The use of such linguistic structures in Ants' speech is a means of concealing the not entirely legal nature of their work and is closely connected with permanent lack of money. The real meaning of the money and 'a carton' is identical: one has it for a moment and then it 'turns into' objects of another kind. Such metaphorical treatment of cigarettes can be noticed even more clearly in analyses of complex economic transactions carried out by Ukrainian Handsellers. They buy for dollars 'a carton', some vodka and a few beers in one-litre, plastic bottles. After losing one dollar in the Ukrainian customs office (bribe) and slightly less thorough check-up in the Polish one, they begin a few-hour standing with the goods near the bars in Medyka. Having successfully sold their products, Ants from Ukraine go shopping in the supermarket 'Biedronka' or in the bazaar and spend all the money they had earned that day. First of all, they buy food: onions, oil, meat, tomatoes, sausages, coffee, fruit, etc. I initially thought that those products are cheaper in Poland than in Ukraine (which is true about onions). Inhabitants of Ukrainian Mościska explained to me the real reason of shopping in Medyka:

> W1: It's cheaper here [in Poland] but for the goods. Cause if you exchange for the money, it is not.
> W2: It's very expensive on our side. I will take 50 hryvnias and I will buy nothing. … And here for these 50 hryvnias I will buy the same products, then will sell them and will buy all that I need.[4]

In this particular group of Ants the language of the local economy has even more explicit 'barter' character, which reflects thinking in terms of 'goods' instead of money.

Constant 'walking' and 'carrying' i.e. laborious movement of thousands of Ants on a circle of the border is presented as the essence of the border itself. In the local discourse the border spins; it constitutes a movement. According to Ants' experience, not only space but also the structure of time affecting the border resembles a circle. The nature of this local time concept is visible when the Ants' discourse is confronted with the geopolitical or bureaucratic discourse of the border.[5] From the geopolitical perspective of an average citizen of Poland the border is analysed in the historical time of a linear structure. As such, it succumbs to the chronological sequence of events taking place in this part of the world. Marked after the war, it used to separate the People's Republic of Poland from the USSR. Then, it separated Poland from Ukraine and presently also constitutes the border of the EU. Poland's joining the Schengen Area is a consequent stage of this border's history.

Ants describe the past of the border using means of expression characteristic to the myth:

> M.B-Sz.: Since when has there been a border in Medyka?
> M: Since I was born. This border crossing has always been here. (...) My friend who lives near the border told me about the times before the war. There was a village, and then they came, destroyed it and made this border crossing. Medyka is here, Szeginie is there. And this was built together before the war and there was no gap.[6]

Here the time has a circular, repetitive, closed structure. As Hastrup puts it, the myth explains the present by recalling 'the time of the beginning', when reality was gaining its qualities (Hastrup, 1997, p. 25). Marked after the Second World

---

4 Women aged around 50, the Poles from Mościska. They work in the farm and as Ants.

5 The discourse of procedures, reports and statistics illustrating successes and failures of the border administration, also the language of international agreements ratification such as Poland's access to the Schengen Area or the agreement on small-distance cross-border movement. In those discourses the essence of the border is its subjectivity and stability. It is marked, administrated and submitted to the decisions of its supervisors.

6 Male aged around 40 - the Poles from Mościska, occasionally carries cigarettes.

War the border became a wall dividing a double ethnic village. After the opening of the borders in the early 1990's the Polish and Ukrainian Ants united this place again to make it function as 'a village'. Consequently, local habits shared by most people regardless of nationality, a specific language and relations based on everyday community practices were created. Poland's joining the Schengen Area and introduction of the new customs regulations in fact result in closing of the border again. Their current tragic situation Ants see as a repetition of a mythical collapse of the primary order caused by alien decision-makers; for them it means the return to the times of 'the Iron Curtain'.

During the Ants' protests the key metaphor which helped define the current situation was the metaphor of the Polish People's Republic and the martial law. Protesters emphasized that: 'the law' came like the state of war, at night' and that the authorities also attacked them in December, just like before. They compared the police to ZOMO[7] and were convinced of being invigilated by agents.[8] The following conversation describes the police intervention aimed at dispersing the Ants who blocked the border the first night:

> W1: It was virtually a lynch. There were twenty police cars, they stood with shields, people were in the middle of the street, they pushed people under the trucks (...) those who fell down were taken to paddy wagons.
> W2: These were shooting at us.
> W1: They were shooting with rubber bullets. It was a lynch. The 1980's. (...) You know, I was a teenager during the martial law period, I'm telling you, I haven't seen anything like that. It was an authentic lynch.
> M: They behaved like ZOMO officers.
> W1: They were taking truncheons, women were falling down, because they were pushing them and hitting with truncheons.[9]

From the local perspective the struggle with the new customs regulations – a responsibility of 'the authorities in Warsaw and Brussels' – is a struggle with an oppressive, violence-based, alien system of power oriented against citizens which resemble the communist system. The protest of Ants who exhort the community to 'solidarity' is in this context a valid solution going along the national tradition.

In such context the immobility of Ants in Medyka absorbs additional meanings and becomes more comprehensible. Metaphorically, the Ants' statements and

7 ZOMO – Zmotoryzowane Oddziały Milicji Obywatelskiej (*Motorized Reserves of the Citizens' Militia)* functioned between 1956 and 1989 and was used to take care of the public order in critical situations, which meant dispersing and pacifying (most often in a very brutal manner) anti-communist demonstrations.

8 During my fieldwork my ID was checked several times because – as explained by one of the protesting men – 'there might be agents'. Unknown people crossing the border were called names: 'informer', 'spy', 'leftist', 'blockhead'.

9 Women aged around 40 and men around 40, all Ants.

actions correspond with the massive labourers' strikes in the 80's and the entire tradition of 'Solidarity'. The border understood as the biggest workplace in the region – the factory – ceased functioning. It stopped the production, i.e. cigarette smuggling which, according to Ants, the whole region's prosperity is based on.

The Border as a Space of Communication

When I was making a selection of the region to conduct my research a few years ago, I chose Medyka among all other border crossings mainly because it was the only place where people are 'a talking community' ('palpable unit', Møhl, 1997, p. 36). In their border crossings dominated by cars, buses and trains, the space of communication is much more limited.

I assume that no other town in Poland owns a similar forum of information exchange which the sinkhole was for Przemyśl. Everyday a few thousand people cross their ways at the border. They spend long hours in the sinkhole waiting for the customs check-up and participate in the constant communication. The rumble in the sinkhole never quietens. The basic information spread among people every few hours is 'whose shift it is', i.e. which customs officer is currently working in the Polish customs office (Ants perfectly recognize all officers – each of them has a different way of conducting the check-up, therefore Ants adjust smuggling techniques to 'who is on shift'). The news, passed with mobile phones, would immediately spread across the entire county. The 'Categories of Talk' (Saville-Troike, 2003, pp. 26–27) specific to the language of the community of my interest could be divided into two main groups: ritual announcements and fabularized narrative statements. First category includes unceasing conversations about cigarettes' prices, settling which customs officer is on duty and how 'he checks up' as well as defining the speed of 'leaving' the sinkhole. Exchange of these statements has no informational character. Ants are very well acquainted with the prices and after a few minutes of being in Medyka they already know which officer is currently 'groping'. However, they can still discuss the same subjects repeatedly. 'How much is the blue Elem today?', 'Bond is stable', 'They don't take the red Marlboros', 'Chesters make no profit anymore'[10] – these ritual announcements keep placing Ants in 'the world of common references' (Rothenbuhler, 2003). This also characterizes talking of politics, complaining about health, telling jokes, singing, and praying together.

The other fundamental category of Ants' speech is making more or less complicated stories: from rumours, through stories of heroic adventures at the border to 'blood-curdling' stories of fatal accidents, murders and rapes at the border. The most popular local themes are stories of spectacular smuggling, ways of cheating customs officers and cases of death in the sinkhole. I also came

10 'Elem', 'Bond', 'Malborek' – these popular brand names of cigarettes are locally used in diminutives (respectively: L&M, Bond, Marlboro)

across certain distinctive narrative plots which turned out to be a base of many stories passed onto me in many different versions such as a story of a local boy killed by a customs soldier, a story of a 'stupid Kazio' (a local freak) or stories about orthodox Jews sometimes crossing the border there. Through the talking of 'others' a complicated mythology is created in the sinkhole and its heroes are Ants, Buyers, custom officers and soldiers (Møhl, 1997).

However, not only the knowledge of specific stories determines the affinity to a symbolic universe of Ants in Medyka, but most of all functioning in the linguistic reality, in which the story exchange is an everyday activity. The fundamental element proving Ants to be a 'speech community' (Saville-Troike, 2003, p. 15) is the local experience of participation in the incessant talk (cf. Møhl, 1997, p. 36).

The Border as the End of the World

Changes introduced to the customs regulations which turned out unfavourable for Ants were treated as brutal interference of the alien authorities. They destroyed the local concept of the region as an independent entity. During the protest I was presented to Podkarpacie (the border district in which Przemyśl is second biggest city) as the worst and most peripheral part of the country. It was interesting to examine the local language of Poland's geography. The division of the country into its better and worse parts was not made along with the forced by the media West–East division, where the East is called Poland B (worse, retarded, deficient, social), while the West is named Poland A (better, modern, European, liberal). Rather, it was made with the help of the following terms: 'upper', 'middle' and 'bottom'. Podkarpacie (called the end of the world, the end of the EU, a hole, Poland D, Ethiopia, the Third World or Iceland due to the economic crisis there) was contrasted with 'the middle Poland', 'upper Poland', with the cities situated 'up there in Poland'. In the word 'Podkarpacie' the prefix 'pod' was strongly emphasized.[11]

In the book *Metaphors We Live By*, George Lakoff and Mark Johnson distinguish and describe so-called *structural metaphors.* Those two are key methods of metaphorical understanding of the entire classes of phenomena in our culture (Lakoff and Johnson, 1988, p. 8). One type of those metaphors is 'orientational metaphors'. Transposition of the entire system of concepts referring the spatial orientation (up–down, in–out, front–back, on–off, deep–shallow, central–peripheral, etc.) into other systems the authors relate to having 'these and no other bodies' which function in space. Having analysed metaphorical examples of using terms 'up' and 'down', they list the following oppositions: happy–broken down, healthy–ill, alive–dead, being in power–submissive, dominating–dominated, much–little, valuable–worthless, good–bad (Ibid., pp. 35–40). The above set of

11 The prefix 'pod' means 'under'. Inhabitants of Podkarpacie underline their inferiority towards the 'upper' and central Poland (translation mine)

binary oppositions perfectly illustrates the way Ants define themselves and their region in comparison to the 'rest of Poland' in the critical moment of changes.

Another metaphor linked with the local geography of Poland is the metaphor of 'closing'. Ants described their region as 'forgotten' and completely 'cut off' the rest of the country. What is interesting, the above mentioned 'rest' was also defined as strictly separated from everything that is 'here'. The below conversation about the way people spend Easter might be a good illustration:

> W1: we are close to our families. We invite each other. It's modest but in the family. I went once to Warsaw for Easter and I'm not going to go back ... There's no Easter there. There's no atmosphere ...
> W2: I was sitting in four walls when I was abroad for Easter.
> W3: Yes. In Warsaw they lock themselves. Everything only for themselves. I don't understand it.
> W1: We are good people here ... [12]

Møhl argues that closing and isolation are in many ways communicative dimensions. In most cases there is no real isolation but people describe their contacts with the outside world in terms of being cut off due to impossibility of a direct exchange and a dialogue (Møhl, 1997, p. 21). The conviction that 'the world outside' differs extremely from 'ours' emerges not due to inaccessibility of this world but because of different communication patterns. Metaphors of being imprisoned 'in four walls' abroad or 'locked doors' of Warsaw flats show inability to communicate due to the lack of common experiencing among which the communication might have been organized.

The Border as Prison

The last two metaphors which I would like to develop – the metaphor of the border-prison and border-front – complete the local image of the border with elements connected with the body, suffering, pain and humiliation. I heard more than once that to understand the border one needs to 'see it with one's own eyes', 'cross it several times', 'wait in the sinkhole for a few hours' or even 'be born here'. To get to know the Ants' border one should practice the border, smell it, hear it and feel the pain it causes.

Violence is inscribed in the very architecture of the border, designed – according to my interlocutors – to inflict pain. This is how they describe the sinkhole – place where they are daily crowded and immobile for hours:

> For example, on the Ukrainian side there is no loo. Well, there is on the Polish, but it is sometimes out of order ... Humiliation. These rails are made to create

12 Women aged 40–50, all Ants.

> crowd on purpose. (Male, approximately 60-year-old, works in a parking lot and as an Ant)
> Well, this is Auschwitz. This is just Auschwitz. … You know, when the crowd charges, many people leave with their intestines broken. (Unemployed female, approximately 40-year-old, Ant)
> Sinkhole is a row of stalls, you know, like the ones for animals. (Unemployed male, approximately 40-year-old, Ant)

In Ants' narratives the border and its primary representation, that is the sinkhole, is a place of oppression – space of coercion and imprisonment, where people, crowded there against their own will, feel their own physiology much stronger. In the sinkhole Ants become 'susceptible bodies' which can undergo 'disciplinary techniques'. 'Disciplinary power' as described by Michel Foucault, works and enslaves by means of situating people in isolated, closed spaces in order to allow the guard to freely control their bodies (Foucault, 1995). In the sinkhole Ants are counted, classified, supervised and punished – just like prisoners of a concentration camp, like cattle.

In such context the word 'Ant' acquires new, negative meaning – people here are like small, meaningless animals which, as one of my interlocutors stated, 'can be smashed with a foot'. In the fragment of their book devoted to the human body at the state border, Hastings Donnan and Thomas Wilson claim that: 'Characteristics attached to some people crossing the border are similar to those attached to the certain kinds of animals.' (2007) According to the authors, such classification (conducted even on the verbal level) allows offending them, ill-treating, justifies the inhuman treatment on the border (Ibid., pp. 179–81).

The border as a system of coercion and humiliation has also another dimension. It reveals in the Foucauldian self-discipline and power internalized in human bodies. Such experience is shared mostly by the Poles from Ukraine, selling cigarettes from hand. Poverty pushes them to trade vodka and cigarettes. They often admitted that they felt humiliated by their profession – mostly referring to the vodka trade which, in their opinion (these were mostly older women), is something dishonouring. Handsellers are the most intimidated group at the border. They are driven away by Buyers and bar owners, forced to hide, estranged in Medyka. Deprived of their own territory, disgraced in their own opinion, they express it non-verbally in the unconditional, uncontrolled gestures. They lower their sight, hide the head, assemble in groups, mince, habitually hide 'flask' and 'cigars' under their clothes while seeing a police car or an ordinary cab. These instinctual gestures show the mechanisms of the disciplinary power which affects individuals in order to create a habit of self-control in them. The power works automatically, like the 'Panopticon'[13] – people, reduced to susceptible bodies and

13 Panopticon is the Benthamite idea of a perfect prison in which all prisoners are subdued to the incessant surveillance. It was supposed to be an architecture of perfect control and re-socialization. The prisoners, convinced of their visibility, start to control

convinced of a permanent supervision and of their visibility, become their own guards (Foucault, 1998, p. 196).

Border as Front

The character of the power relations at the border is encapsulated in the very important metaphor of the border-front. Before I understood the meaning of this metaphor, I had had no doubts that the division of power was unambiguous – Ants had been subject to the power of the border administration (customs officers and soldiers). Customs officers are Ants' greatest enemies. In the opinion of my interlocutors, their work is to humiliate Ants, show them domination and contempt. The tools of the customs officer's power are the hands, which 'grope' people as well as various 'extensions' of the hands such as instrument which detects tinfoil in cigarettes. During humiliating search, Polish customs officer can touch, check and give orders. He has a variety of ways to eliminate the Ant from the border. He can maliciously put the stamp across the page, to make the passport expire sooner. He can say 'good night' – in such situation one cannot return to the border during this shift. He can 'give shoe' – make somebody withdraw in Ukraine, or – for greater offence – 'give a bear', a one-year ban to cross the border. However, the Ukrainian check-up is easy – one has merely to bribe an officer. Yet, the Ukrainian border soldiers are, according to the Ants, similar to the prison guards, using the military modes of humiliating people. Customs check-up, personal search, the soldier's drill – it is the Foucauldian 'politics of coercion', reducing people to the army of susceptible bodies. The subject of the power understood in such terms is not any individual customs officer or the border soldier. As Foucault shows, modern power is not executed by a 'sovereign' (a specific personalized ruler), but instituted in numerous procedures the subject of which is the disciplined population. The aim of the authorities is not to protect the territory, but to create and control the disciplined population. Accepting this theory of power, Ants begin to appear as a population which undergoes permanent supervision, simultaneously created and controlled by the customs, passport and penal procedures. Therefore, Ants lose their subjectivity and ability to cause things because – as Foucault writes – the population cannot turn against the procedure which creates it (Foucault, 2000, p. 184).

My understanding of the power relations changed when I grasped the meaning of the border-front metaphor. Border is not a place where power is held absolutely by one party – i.e. the administration. It is rather a 'disputed territory', the area of the incessant struggle between customs officers and Ants. As one of them put it, 'So many years after the war, and here we are on the fucking front' (female Ant aged 20, works in a bar).

---

their own behaviour by themselves. Michel Foucault turned it into a metaphor of modern power and the society under surveillance (Foucault, 1998, p. 208).

For many years of their work at the border Ants have created a system of effective resistance against the authorities. The methods of protest used to fight the customs law and individual customs officers are the strategies typical to 'weapons of the weak' and to 'hidden transcript' – modes of struggle characteristic of excluded groups, described by James Scott. These are different modes of expression: gestures, speech and other 'everyday forms of resistance' (Scott, 1992, pp. 62–5, 1990, pp. 198–200).

At the border I have observed many acts of this type. Ants demand that customs officers give reports from the check-ups, denounce the abuse to the authorities of the office, sue the officers, protest, block the passage; I have even heard about the idea of establishing the petty smugglers' trade unions. Popular forms of illegal resistance are running through the customs office without a check-up, cutting holes in the net which divides Polish and Ukrainian sides of the border, bribing, petty smuggling, participating in the alternative circulation of knowledge, i.e. gossiping and laughing at officers, making offensive information public on the Web, deriding the very customs procedure.[14]

One of the most important strategies used by the Ants to offend and humiliate the customs officers is to give them offensive nicknames. The stigmatizing nickname – widely known – is a sort of punishment by which the officer loses his name and surname in the society. He receives a new, reduced identity. The nicknames are related to the looks, the way of conducting check-ups or the previous profession, e.g.: the Minor, Big Bertha, Little Bitch, Rag, the Tousled, Barbie, Gypsy, Make-Up, Ded Moroz, Grandpa Ironbar, the Storekeeper, Shrew ('grubs' thoroughly), Faggot (he paws), Sable (checks 'hard'), Chicken (he used to sell roasted chicken), Lawyer (used to study law), Smuggler (used to be an Ant).

The border is a front, a place of everyday power-play. Przemyśl, however, is presented by my interlocutors as the territory of Ants, where customs officers are forced to hide. Outside the office, deprived of the stage and regalia of power, they become powerless and susceptible to the attacks of the inhabitants.

Conclusion

In this chapter, I presented a local discourse of Ants and the way they experience the border. By analysing the most popular metaphors of the border, which lives and dies, is a circle, a movement and a factory, a place of meetings, imprisonment and struggle, I intended to show the complexity of the local experiencing of the border. The multiplicity and diversity of meanings inscribed in the concept of the

14 At the border there is a popular story of Stupid Kazio (local fool) who presented excrement to the customs officer. The whole joke relies on the ambiguity of the word 'gówno' (shit), which is also colloquially used for 'nothing'. When the lady officer asked Kazio if he had anything to declare, he answered truthfully. The suspicious officer searched his bag and to her surprise …

border, seemed at first absurd, illogical, internally contradictory and not adding to its coherent image, but turned out to be the most valuable element. Metaphor is treated here not only as a way of perceiving and giving sense to life or putting the narrative about it in a certain order, but also as an interpersonal relation, an occasion for an encounter, an area being 'extension of reality' – mine and theirs (Hastrup, 2008 [1995], p. 25).

The metaphors of the border I have selected obviously do not give a complete picture of the entire local complexity of the concept. The underlying idea of this text and the reflections presented is a hermeneutic conviction that 'the world cannot be understood directly [as a whole]; one can draw conclusions on the basis of its parts and those parts have to be conceptually and perceptually rationed out from the stream of experience' (Clifford, 2000, p 47, translation by the author). The anthropologist's task is to take out significant elements from the fluid life situation giving them reference in a certain context. This way there emerge 'areas of synecdoche whose parts refer to entities which consist in what we call culture' (Clifford, 2000, p. 47, translation by the author). Let the metaphors presented above stand for those parts.

References

Byrska – Szklarczyk, M. (2010), 'Granica jest po to, żeby chodzić – krajobraz kulturowy Mrówek z Medyki', in M. Zowczak (ed.), *Na pograniczu „nowej Europe*', Warsawa: Polsko – ukraińskie sąsiedztwo.

Cegliński, Ł. (2005), 'Społeczność podzielona. Granica litewsko – białoruska z perspektywy procesów komunikowania się', in A. Perzanowski (ed.), *Centrum na peryferiach. Monografia społeczności lokalnej Ejszyszek i okolic na Wileńszczyźnie*, Warszawa: Wydawn. Uniwersytetu Warszawskiego.

Chlebicka, A. (2010), 'Rola przemytu w życiu młodzieży pogranicza. Na podstawie badań w Kalinkowie (pow. przemyski)', in M. Zowczak (ed.), *Na pograniczu 'nowej Europy'. Polsko – ukraińskie sąsiedztwo*, Warszawa: Wydawnictwo Trio.

Clifford, J. (2000), *Kłopoty z kulturą*, Warszawa, Wydawnictwo KR.

Cohen, A.P. (1985), *Symbolic Construction of Community*, New York and London: Taylor and Francis Ltd.

Delmanowicz, D. (2008a), 'Nowy terminal w Medyce', *Nowiny 24*, 05.04.2008 http://www.nowiny24.pl/apps/pbcs.dll/article?AID=/20080605/PRZEMYSL/873081911.

Delmanowicz, D. (2008b), '"Mrówki" grożą blokadą przejścia granicznego w Medyce', *Nowiny 24*, 26.11.2008 http://www.nowiny24.pl/apps/pbcs.dll/article?AID=/20081126/PRZEMYSL/305011608.

Donnan, H. and Wilson T.M. (2001), *Borders, Frontiers of Identity, Nation and State,* Oxford: Berg.

Gołębiewski, J. (2010), 'Drobny handel wahadłowy na pograniczu polsko – ukraińskim w ujęciu antropologicznym', in M. Zowczak (ed.) *Na pograniczu 'nowej Europy'. Polsko – ukraińskie sąsiedztwo*, Warszawa: Wydawnictwo Trio.

Gołębiowski, P., Krassowska, K., Strzemińska, A. and Śpiewak, J. (2009), 'Granica i przejście graniczne w perspektywie aktorów lokalnych', in W. *Łukowski*, H. Bojar, B. Jałowiecki (eds.) *Społeczność na granicy. Zasoby mikroregionu Gołdap i mechanizmy ich wykorzystywania*, Warszawa: Wydawnictwo Scholar, pp. 137–64.

Jagnyziak, K. (2010), 'Wspólnota interesów czy rodzinny interes? Wpływ handlu przygranicznego na relacje w rodzinie', in M. Zowczak (ed.) *Na pograniczu 'nowej Europy'. Polsko – ukraińskie sąsiedztwo*, Warszawa: Wydawnictwo Trio.

Foucault, M. (1995), *Discipline and Punish. The Birth of the Prison,* London and New York: Penguin.

Foucault, M. (1991), 'Governmentality', in G. Burchell, C. Gordon and P. Miller (eds.), *The Foucault Effect: Studies in Governmentality*, Chicago, IL: University of Chicago Press, pp. 87–104.

Hann, C. and Hann, I. (1992), 'Samovars and Sex on Turkey's Russian Markets', *Anthropology Today* 8(4): 3–6.

Hastrup, K. (1997), 'Przedstawienie przeszłości. Uwagi na temat mitu i historii', *Polska Sztuka Ludowa – Konteksty*, t. LI, z. 1–2: 22–7.

Hastrup, K. (2008 [1995]), *A Passage to Anthropology: Between Experience and Theory*, London: Routledge.

Henrik, E. (2006), 'Cross-border Small-Scale Trading in South-Eastern Europe: Do Embeddedness and Social Capital Explain Enough?', *International Journal of Urban and Regional Research* 30(2): 346–61.

Johnson, M. (1987), *The Body in the Mind. The Bodily Basis of Meaning, Imagination and Reason*, Chicago, IL: University of Chicago Press.

Kazimierska, K. (2006), 'The Border as an Individual and Collective Experience – The Sejny Case', in *Borders and Borderland*, Warszawa: IFIS PAN.

Konstantinov, Y. (1996), 'Patterns of Reinterpretation: Trader-Tourism in the Balans (Bulgaria) as a Picaresque Metaphorical Enactment of Post-Totalitarism', *American Ethnologist* 23(4): 762–82.

Konstantinov, Y. (1998), 'Outclassed by Former Outcasts: Petty Trading in Varna', *American Ethnologist* 25(4): 729–45.

Lakoff, G. and Johnson, M. (1988), *Metafory w naszym* życiu, Warszawa: Aletheia.

Matejko, E. (2008), 'Przejście graniczne jako zasób społeczności lokalnej', in H. Bojar, D. Wojakowski and A. Sadowski (eds), *Pogranicze. Studia społeczne*, t. XIV, Białystok, pp. 61–77.

Møhl, P. (1997), *Village Voices. Coexistence and Communication in a Rural Community in Central France*, Copenhagen: Museum Tusculanum Press.

Plińska, W. (2010), 'Turystyka handlowa na południowym Podlasiu – perspektywa *emic*', in M. Zowczak (ed.) *Na pograniczu 'nowej Europy'. Polsko – ukraińskie sąsiedztwo*, Warszawa: Wydawnictwo Trio.

Rakowski, T. (2009),Łowcy, *zbieracze, praktycy niemocy. Etnografia człowieka zdegradowanego*. Gdańsk: Sopot.

Rapport, N. and Overing, J. (2000), *Social and Cultural Anthropology: The Key Concepts*, London: Routledge.
Ricoeur, P. (1971), 'The Model of the Text: Meaningful Action Considered as Text', *Social Research* 38: 529–62.
Ricoeur, P. (1976), *Interpretation Theory: Discourse and the Surplus of Meaning*, Fort Worth: Texas Christian University Press.
Ricoeur, P. (1989), *Język, tekst, interpretacja*, Warszawa: Biblioteka Myśli Współczesnej.
Rothenbuhler, E. (2003), *Komunikacja rytualna. Od rozmowy codziennej do ceremonii medialnej*, Warszawa.
Saville-Troike, M. (2003), *The Ethnography of Communication. An Introduction*, Oxford: Oxford University Press.
Scott, J.C. (1985), *Weapons of the Weak: Everyday Forms of Peasant Resistance*, New Haven, CT: Hidden Transcript.
Scott, J.C. (1990), *Domination and the Arts of Resistance*, New Haven, London: Hidden Transcript.
Scott, J.C., (1992), Domination, Acting and Fantasy', in C. Nordstrom and J. Martin (eds.), *The Paths to Domination, Resistance, and Terror*, Berkeley, Los Angeles, Oxford: Oxford University Press.
Wąsowska, E. (2010), 'Adaptacja ekonomiczna mieszkańców Stariawy do zmian politycznych zachodzących na granicy polsko – ukraińskiej od lat 50. XX wieku do czasów współczesnych', in M. Zowczak (ed.) *Na pograniczu 'nowej Europy'. Polsko – ukraińskie sąsiedztwo*, Warszawa: Wydawnictwo Trio.
Zowczak, M. (2010), 'Na pograniczu „nowej Europy", Polsko – ukraińskie sąsiedztwo, Warszawa: DiG, Instytut Etnologii i Antropologii Kulturowej.

Chapter 6
Border Orderings: The Co-existence of Border Focusing and Letting Border Issues take the Back Seat at the German–Polish Border

Marie Sandberg

Introduction

Among border scholars it has been widely acknowledged that in order to better grasp and understand the de- and re-bordering processes of contemporary Europe, attention needs to be paid to the many different ways borders are created, shifted, and transformed in practices of everyday life. As argued in the Introduction to the present volume, this turn to practice is signalled by central concepts such as 'borderwork' by Rumford (2006; 2008); suggesting that a whole range of actors need to be taken into account in explaining current and practice-driven transformations of the borders around the globe. Another example is the concept of 'bordering' as elaborated by van Houtum and van Naerssen (2002) which conceives borders as social practices of spatial differentiation that 'make a difference in space among the movements of people, money or products' (ibid, p. 126).

As a consequence of this turn to practice within border studies, focus is directed towards actors of everyday life (cf. Andersen in this volume; Galasińska et al., 2002; Linde-Laursen, 2010; Löfgren, 2008; Sandberg, 2009a, 2009b). These everyday life actors are, along with geopolitical actors such as states, empires, and maybe even more diffused forces such as 'globalization', regarded as contributors to the ongoing making and remaking of international borders. Thus, one implication of practice-oriented border research is that agency is distributed among several actors, including those of everyday life.

Yet, how do we conceptually work with this rich variety of actors within border research?

By introducing the notion of 'modes of ordering' to border studies this article suggests a further opening of the floor not only for actors of everyday life but for yet a broader and more analytically sturdy concept of what constitutes an actor; to put it quite simply there may be even more actors relevant to investigate in the de- and re-bordering processes of Europe than we thought. Further, the notion of modes of ordering can introduce the idea and grasp that there are always more than one mode of ordering at stake when analysing how borders are formed in practice. Borders are thus conceived as the practical effects or products of the ongoing work of contingent and recursive modes of ordering.

With the modes of ordering being developed by sociologist and scholar of science and technology studies John Law, the argument is built on performative approaches from the field of Actor Network Theory (ANT). These approaches imply an idea of delegated agency where agency is seen as an outcome of relations between humans and materialities, as well as a focus on how realities (in plural) are in a constant process of becoming through ongoing and recursive enactments (Mol, 1999, 2002; Law, 1994). Further, it is a notion based on a principle of analytical symmetry between knowledge practices. This means that all knowledge arises from specific cultures of science practices and can thus be considered situated and practice based (Law, 1994, 2009).[1] Hence, modes of ordering rely on a material-semiotic premise that entities are translated into being, as they achieve agency through relations to other entities in heterogeneous networks (Callon and Law, 1997; Latour, 1996).

So, how might the modes of ordering add to the practice turn of border studies? I will explore this through an analysis of how cross-border integration processes are handled and organized in the cities of Görlitz and Zgorzelec, situated on either side of the German–Polish border. Since the early 1990s it has been a goal for these two cities to enhance integration and cooperation across the historically burdened Oder–Neisse border. With this goal follows the strategy of also decreasing the position as a peripheral border area with severe structural problems such as deindustrialization, unemployment and migration from the area (Hanneman et al., 2000). Due to the fact that the younger population is migrating from the border area, the youth is a target for the cross-border integration in Görlitz and Zgorzelec. However, as became evident during my field research in Görlitz and Zgorzelec in 2005 and 2007 respectively, these youth tend not to enrol in the cross-border activities offered. It is therefore of particular interest how the local cities' leadership such as the city councils of Görlitz and Zgorzelec attempt to mobilize the youth to engage in the cross-border activities and the transboundary identity projects.

A Cross-border Regime

A common denominator for the main part of the modes of ordering to be presented here is concerned with how to create passages, or even how to transgress, the German–Polish border as a barrier. The context for these particular border practices can be characterized as a cross-border regime (Kramsch and Hooper, 2004); which makes it interesting to focus on the practices – namely those of the youth – that

---

1 The principle of analytical symmetry within Actor Network Theory derives from the tradition of the Sociology of Scientific Knowledge (SSK). I will not elaborate further on this theoretical foundation, however, it was within this realm that Bloor formulated the SSK premise that all knowledge practices are symmetrical and thus deserve to be explained and described in the same terms (cf. Law, 1994, 2009).

seem not to fit in. It is on this backdrop that the analysis shows how such 'unfitting' modes of ordering occur along with those modes of ordering that, through the means of projects funded and supported by municipal, regional, state, and EU levels, continuously aim at crossing the border. Paradoxically, then, this aim of breaking down the border as a barrier also includes a constant focusing on the border. This endeavour butts, yet again, up against those practices of the youth who rather feel like letting the border take the back seat.

Outline

The first section introduces the notion of modes of ordering and how it implies an idea of delegated agency. The second section introduces the German–Polish border history and the various cross-border integration projects of Görlitz and Zgorzelec. In the third section I present and develop three modes of ordering applicable to the German–Polish case: Enterprise, Administration and Allocation. By describing the different ways in which the German–Polish border is put into focus as well as (temporarily) ignored, I attempt to analyse how the border is enacted differently and how this multiplicity of 'border orderings' co-exist. In conclusion I discuss how these interrelated modes of ordering can be characterized with a concept derived by the Dutch philosopher Annemarie Mol as a 'living-in-tension' (Mol 1999, 2002).

The Modes of Ordering

The notion of modes of ordering was initially developed and presented in the book *Organizing Modernity* (Law, 1994). The book is based on a one year fieldwork conducted by John Law at the Daresbury research laboratory, UK, specialized in experimental science and engineering (Ibid.; cf. Law and Moser, 1999). Whereas the modes of ordering is a novel approach within border studies (Sandberg, 2009a, 2009b), it has been applied and challenged in a wide range of other empirical fields such as organization studies (Jensen, 2001, 2004), health care practices (Mol, 2002, 2008; Moser, 2005), and practices of public administration and management (Petersen, 2011), to name a few.

Law introduces the modes of ordering in three steps. Firstly, rather than attempting to explain 'the social order', which has been one of the main concerns within the classical sociological discipline, the modes of ordering depict orders in plural. Secondly, Law stresses that the modes of ordering are to be understood as verbs rather than nouns to avoid essentializing them and to stress their character of being in a constant state of becoming. Thirdly, ordering is understood as materially and discursively heterogeneous. This means that the notion of modes of ordering aims at grasping patterns in the different ways relations between people and stuff are organized. The modes of ordering are thus defined as describing 'recurring patterns embodied within, witnessed by, generated in and reproduced as part of the ordering of human and non-human relations' (Law, 1994, p. 83). The modes of ordering do have strong family resemblances with Foucault's concept of discourse, as they, too, can be seen as patterns of ordering within a relatively structured whole. However, understanding this 'whole' along the line of a Foucauldian epistéme, will according to Law turn the orderings into a too coherent pattern of order (in singularis) or other kinds of 'big blocks' (Law, 2000, p. 27), which is what Law tries to avoid (cf. Mol, 2002, p. 62). This might also be the reason why the list of modes of ordering is stressed as being neither complete nor exhaustive (Law 1994, p. 83).

In the following I will discuss how the three features mentioned above: from order to ordering, ordering as a verb rather than a noun, and ordering as heterogeneous discursive-material, implies a concept of delegated agency, and how this applies to a practice oriented border study.

Delegated Agency

Taking the step from investigating 'the social order' to order*ing* implies that from the outset, no privilege is given to certain practices; everything is regarded as outcomes of ordering (in plural). The specific modes of ordering of Law's study of the research laboratory are developed through the readings of political and theoretical discourses as well as empirical material, such as observations among and conversations with his interlocutors. Hence the modes of ordering perform an analytical symmetry between political, economic, cultural, and scientific practices (Law, 1994, pp. 9ff, 12). Importantly, the principle of analytical symmetry does not mean that the 'world is flat' or that everybody is considered to be equal. Rather, the potential of this approach, as I see it, is to analyse and unfold how differences and the strength of some practices over others do come about and stabilize, as they are products or effects of ordering.

Depicting 'ordering' as a verb, and not a noun, also means that purification and homogenization (in a negative sense) are not the only outcomes. Hence, it can be argued, that modes of ordering turn processes of ordering away from a more critical tradition of linking ordering and 'Othering' (cf. Houtum and Naerssen, 2002). Ordering can instead be seen as an ongoing tension between many different logics and rationales, each trying to ensure their effort.

In applying the modes of ordering as suggested here, ordering is understood as materially and discursively heterogeneous which means that the patterns that constitute them encompasses ways of organizing relations between non-humans as well as humans. Ordering is thus depicted as a multiple endeavour encompassing practices of many kinds; be it those of cross-border cooperation at municipal level, EU integration policies, national level regulations, or everyday life practices (cf. Law, 1994, p. 188). To this list other socio-material actors relevant for border practices could be added; such as customs houses, bridges, and policy documents.

Due to their performative character no specific actors or personality types are attached to the modes of ordering such as the 'rule followers' versus the 'entrepreneurial' type of person. The modes of ordering are each generating specific actor traits that can be seen as effects of their particular ways of ordering the world

(Law, 1994, p. 74). It is in this sense, that the modes of ordering distribute agency to several discursive-material actors.

Thus the modes of ordering combine the concern for processes of bordering as social practices and leaves open the possibility for acknowledging the role of potentially surprising actors, who also take part in the ongoing transformation of the borders of Europe.

In the following I exemplify how the modes of ordering can be applied to a study of cross-border integration processes at the German–Polish border.

The German–Polish Border

Görlitz and Zgorzelec were, pre-1945, one German town (Görlitz) crossing the river Neisse. However, after World War II Görlitz was divided by the new 'Oder–Neisse line', between Germany and Poland, that eventually came to follow the rivers of Neisse and Oder (Schultz, 2001, 2002). Expulsions of German and Polish populations followed this new border demarcation; approximately seven million Germans were expelled from the areas of Eastern Prussia, Pomerania, and Silesia whereas around two million Poles, forcibly removed from now former Eastern Poland, were resettled in the new Polish Eastern territories on the Oder Neisse border (Fure, 2001). During socialist times the relation between the German Democratic Republic (GDR) and the Polish People's Republic (PPR) has been characterized as tense (Schultz, 2002). The fact that the GDR and the PPR were pronounced so called 'brotherhood states' did not result in strong ties or well consolidated communication practices across the border; quite to the contrary (Jajeśniak-Quast and Stokłosa, 2000). During the Cold War the GDR and the PPR turned their backs against one another, manifested through strict control at the border crossing points. With the exception of a limited period from 1972—1980, visas were compulsory in order to cross the border (Schultz, 2002). Hence, the reorganization of territories and populations resulted in a tense post-war German–Polish relationship that contributed to making cross-border integration a special area of attention within the last two decades. This integration has to a great extent been enhanced by EU cohesion programmes and frameworks for regional cooperation. After 1990 the German–Polish border was allotted a pioneering role with the expectation that German–Polish relations could be improved through intensive border-crossing cooperation. Likewise Poland's accession to the EU in 2004 and the Schengen Agreement in 2007 were followed with the expectations that Europe provides the best possible common and future ground for cooperation.

The 'Europa Stadt'

There are several examples indicating that one of the local goals of Görlitz and Zgorzelec is to turn them into assets in the European integration processes (Heimpold, 2001). As a symbolic manifestation of the mutual relationship and cross-border cooperation the two cities pronounced themselves to be one 'Europa Stadt [Europe City]' in 1998. This resulted in a shared effort to become European cultural capital 2010 using the slogan 'Wir bauen Europas Kulturhauptstadt [We build the European Cultural Capital] – From the Middle of Nowhere to the Heart of Europe' (cf. Becker 2005, Grossman 2006). This collaborative effort did not succeed in winning the competition, though. With financial support from the EU, the land of Saxony and the municipalities of both Görlitz and Zgorzelec, the pre-war thoroughfare was re-established in 2004 by a pedestrian bridge across the river Neisse/Nysa.

The cross-border cooperation 'Euroregion Neisse–Nisa–Nysa', which also includes Czech areas, provides yet another example. This Euroregion was established in 1991 and as a framework it provides a means of attracting business along with developing tourism and cultural heritage projects.

Figure 6.1 Pedestrian bridge across the river Neisse/Nysa

Hence, Görlitz and Zgorzelec prove an especially suitable area for analysing the multiple orderings of the German–Polish border, historically and in present projects and policies.

The Migrating Youth

Within the EU regional funding frameworks both Saxony and Lower Silesia, of which Görlitz and Zgorzelec respectively are a part, are categorized as convergence regions with a GDP below 75% of the average level among EU member states. Within the German–Polish border region brown coal mining still exists, however, recently production has decreased due to the fact that within the next 20—30 years the coal deposits will be depleted (Regionaler Planungsverband Oberlausitz-Niederschlesien, 2011). The deindustrialization is accompanied by high rates of unemployment and migration which is seen as characteristic for several smaller and medium sized cities, towns, and regions of the periphery of the former socialist states of the GDR and the PPR (Hannemann et. al., 2000). In 2006/2007 Görlitz counted app. 58,000 inhabitants and Zgorzelec around 35,000 inhabitants (Stadtverwaltung Görlitz, 2006/2007; GUS, 2007). Compared to the population numbers, in the year 1990, of 72,237 in Görlitz this means a decrease in population of 19.7 per cent since the end of the GDR (Schultz, 2002, p. 161, table 1; cf. Becker, 2005, p. 177). Hence, Görlitz and Zgorzelec inscribe themselves into the category of 'shrinking towns' of the German–Polish border area (Hanneman, 2003; Schultz, 2002) and Görlitz is especially in danger of becoming a 'Seniorenpolis', i.e. a 'town for the elderly' (cf. Becker, 2003). The 2005 unemployment rate in Görlitz was 26.7 per cent (Stadtverwaltung Görlitz, 2006/2007), and the 2006 unemployment rate in Zgorzelec was 17.20 per cent (GUS 2007), including youth unemployment as a growing problem. With these facts follows great concern as to how to encourage the youth of the border area to stay and to be engaged in their local environment. This is seen in everyday life discourse and often discussed in the local media.

The participation in cross-border activities appears to be one of the means used in order to enhance this engagement; and is also the reason why I pay special attention to the relation between this particular group of activities and the young people.

In applying the notion of modes of ordering the following analysis explores how there are certain patterns in the way the cross-border integration is practiced differently in the case of Görlitz and Zgorzelec. I will show how the patterns described by this concept can be imputed to the German–Polish border practices, which in this case encompasses the activities of municipal coordinators of and representatives for cross-border cooperation, along with border related activities of the youth living in the two cities at the border.

First a few words on how I developed the specific modes of ordering in order to apply them to the particular ethnographic material of the analysis.

Three Modes of Ordering: Enterprise, Administration and Allocation

The modes of ordering that I wish to elaborate here are Enterprise, Administration and Allocation.[2] I would first like to stress that they are ideally typical in character, hence one would never see them realized one-to-one in practice. Thus they are neither suggested to be representative of all possible ways of ordering the German–Polish border, nor are they meant to be exhaustive (Law, 2004, p. 2). The notion of 'mode of ordering' is an analytical tool imputable to the patterns of the relational activities of socio-material entities and their way of co-existing with each other. Hence, they enable analysis of the multiplicity of how the German–Polish border is enacted.

The ethnographic material consists of observations, the use of local media such as newspapers and policy documents, together with interviews with local and municipal administrations, museums and cultural institutions, as well as with one group of German and Polish high-school pupils. In order to allow the effects of socio-material actors to be taken into account the ethnographic research design included the drawing of maps by the pupils and, what I have termed, 'walking conversations' undertaken with these pupils around the cities of Görlitz and Zgorzelec (cf. Sandberg, 2009a, 2009b).[3]

I will present and apply the three modes of ordering in two steps: first I will introduce the modes of ordering in a rather condensed version highlighting three features I found most relevant when applied to my material: the (1) *virtues* and specific (2) *means* that each mode of ordering is performing, and the (3) kind of *ideal actor* (proper person) they each seem to generate (cf. Law 1994, p. 110ff; Law and Moser, 1999, p. 5). This presentation serves to illustrate how the three modes of ordering are developed with a clear anchoring in the empirical material; such as interview material and policy documents, along with theoretical insights and resources. The second step of analysis will, by a more empirically rich story, exemplify how these modes of ordering co-exist.

Enterprise –the 'Pushing' of Others to Perform Well

As a mode of ordering, Enterprise generates a means for actions such as innovation, i.e. the constant search for new resources and possibilities. This is attempted by using and optimizing the available structures, and is often sought achieved by rule bending and risk taking (means). One important virtue generated by the logic of Enterprise is the monitoring of the actions of others in order for them to perform

2 In the study of Daresbury Laboratory, Law develops further the modes of ordering 'vocation' and 'vision'. However, while the latter refers to different managing styles of the laboratory and the first mentioned refers to a field of experts I find them not to be applicable in this empirical context (Law, 1994).

3 For an analysis of the maps of activities made by the German and Polish pupils, see Sandberg 2009a, 2009b.

well (Law, 1994, p. 176). Thus, one can say that within this mode of ordering, actions are assumed as outgoing 'from below', or from the 'entrepreneurs' themselves, yet these are capable of 'pushing' others into action. Hence the ideal actor that Enterprise effects, is the pro-active, innovative entrepreneur that 'makes things happen'.

In elaborating Enterprise as a mode of ordering Law is inspired by ideals from liberal market theorists such as the virtue of accepting greater responsibility for oneself and one's actions.[4] Also, and according to the principle of analytical symmetry, Law draws inspiration from differentiations made by his interlocutors such as between risk taking 'cowboys' and the rule-abiding 'civil servants' (Law, 1994, p. 69).

There are several examples of how Enterprise as a mode of ordering is present within the border practices that unfold in Görlitz and Zgorzelec. In this section I will provide one illustrative example. During my fieldwork in 2007 I discussed future possibilities for further development of Görlitz and Zgorzelec with the Polish mayor of Zgorzelec. The mayor told me about plans for a high-tech scientific and industrial park in the neighbouring city of Bogatynia which had the potential to boost the area:

> We are trying to develop the city, or the whole municipality. I don't want to say anything too early yet, but it is 99% sure that Siemens is going to open an agency here, there is going to be a design office, and we will also try to attract engineers … It is a great success for me, because we attract well-educated people with a high income. To attract that kind of people is a success for the city. These people are also most likely young. (Interview20070525a).

This excerpt shows that the development plans of the Zgorzelec mayor aimed at attracting a specific kind of 'people'; in this case the high-income, well-educated, and creative types, such as engineers and designers, who would also qualify by being young. It can be argued that this type of plan mobilizes the logic of Enterprise as the characteristics, which are underlined as positive, go hand-in-hand with the actor virtues of the innovative entrepreneur. Creating links between scientific development, design innovation, and production as is the case with this high-tech park is regarded a 'great success' because it will potentially attract new knowledge capital along with increased production possibilities to the area.

In addition, I asked the Zgorzelec mayor how he found the cross-border cooperation to be organized in the best way possible. The mayor did not, as I would have suspected, respond that the municipality should provide the right cross-border

4 According to a further elaboration on the modes of ordering carried out by Law and Moser there is, in contrast to the other modes of ordering, no specific theoretical 'father' behind the mode of ordering of Enterprise; hence the general theoretical inspiration is found in liberal market theorists (Law and Moser, 1999, p. 5).

frameworks or structures in order to secure that the German–Polish integration would find its place; quite to the contrary, he explained as follows:

> Those who want to integrate are integrating … well, there are lots of possibilities, a flood of possibilities … Whatever one might invent, it is also realizable, because on the other side [of the border] there will be someone who will be interested in the project. On the German side there are organizations who are interested in cooperation with the Polish side. (Ibid.)

According to this excerpt the central issue for the Polish mayor was that the integration initiatives derive 'from below', i.e. from voluntary organizations and from the local people themselves. It is a vision that relies on the virtue of the free initiative and that is confident in personal initiatives as an inexhaustible resource. Thus, this presupposition of the proactive and innovative types of actors that are capable of establishing cross-border initiatives is performing a version of the German–Polish cross-border cooperation in line with the traits of the Enterprise mode of ordering. For this kind of logic, it is a matter of – not a structuring from above – facilitating business opportunities, and thus attracting the right kinds, of innovators and entrepreneurs in order to make cross-border integration function well.

Administration – Securing Due Process

Administration performs the bureaucratic way of organizing. This is done by dutiful work according to rules given, i.e. through instructions by superiors (means) thus securing due process (virtue). In order for this mode of ordering to perform well (ideal actor) the demands given and resources provided need to be proper. This can be achieved through the making of clearly formulated agendas. The clear communication of such agendas can be secured through the systematic writing of minutes, the appropriate documentation, and through the follow-up on decisions and, subsequently, their implementation (cf. Law and Moser, 1999). This mode of ordering resembles, to a high degree, the well-known bureaucratic logics of Western protestant societies and is also evidently theoretically resourced by classical sociological studies carried out by Weber.[5]

One illustration of how Administration as a mode of ordering unfolds in the German–Polish case is seen in the works of the German–Polish Joint Coordination Committee (JCC) for cross-border cooperation in Görlitz and Zgorzelec. Since its

5 One very important difference between Law's mode of ordering Administration and Weber's works on the bureaucratic and protestant logic is that the latter's infamous Iron cage thesis proposes a slow but nevertheless steady and one-directional process towards an order of Rationality. In contrast to this, Law states: 'I'm not as pessimistic as Weber … For I don't think that orderings ever turn into orders though they may commit unspeakable crimes as they try to do so' (Law, 1994, p. 79).

establishment by the end of the 90s the mayors of Görlitz and Zgorzelec respectively have been meeting with their staff once a month. Core areas of the cross-border cooperation have so far been city restoration and reconstruction, the establishment of a shared strategy for natural disasters and rescue, cross-border infrastructure, youth cooperation, culture, and tourism (Formuszewicz, 2005).

During my fieldwork in 2007 I met with a representative for the mayor of culture in Görlitz (hereafter the representative). The representative described to me how the cooperation works and she also wanted to stress that overall there were no large obstacles within the work carried out by the JCC coalition. Yet, she admitted that there were some language issues. Further, as the conversation moved on, other concerns were brought up. Whilst not perceived as grave obstacles, the language issues, did however seem to have a noticeable impact on the progress of the JCC. According to the representative, the same thing can be understood quite differently by the different actors, for instance whether an agreement is considered to be binding or whether it is considered to be without obligation. Further, the representative described how the same issue could be put on the agenda several times, even though to her knowledge a decision had already been taken during the last meeting. The representative further exemplified this problem with the fact that the city council of Görlitz had established a part time position, in the form of a cross-border youth coordinator situated on the German side of the border. The task of this coordinator is to educate and inform the youth residing on both sides of the border about Europe and cross-border projects (Europa-Haus, 2011). According to the representative, the plan of the JCC was to establish a similar cross-border youth coordinator on the Polish side of the border. However, this had not happened yet, apparently in spite of several requests posed at the JCC meetings. As she formulates it: 'this is a really serious problem; there simply isn't any such contact person in Zgorzelec' (Interview20070626).

I suggest that this excerpt from my conversation with the JCC representative illustrates how the logic of Administration is also at stake in the German–Polish cross-border integration process. If decision-making becomes unclear or there is no follow-up on the agendas and/or minutes, it is regarded as a dysfunctional way of organizing the common cooperative work of the JCC across the border. In contrast to the Enterprise mode of ordering where rule bending or letting things grow 'from below' can be seen as one of the ways to make things happen, the Administration mode of ordering ends up with potentially frustrating situations as is the case when there is a lack of symmetry between the cooperative functions across the border.

Allocation – Patterns of Indifference and Concern

I will introduce 'Allocation' as a mode of ordering with a quote from the *International Herald Tribune* written immediately after Poland's accession to the Schengen Agreement:

> Słubice, Poland: As of midnight Thursday the once contentious border between Poland and Germany will be thrown open. For the most part it has been more a whimper than a bang for the fall of one of the most historically fraught and violently fought over frontiers on earth. Travelling along the 450-odd kilometres, or about 280 miles, of the border … what is most striking is the relative indifference along the way to the change … To the extent that the fall of the border has been followed in the German media, it has been largely scare stories installing metal shutters, putting up barbed wires and even buying guns. But talking to the locals … the fortifiers seem to be a vocal minority. (*International Herald Tribune*, 2007)

The journalist depicts here the contrast between on the one hand the expectations, foremost generated by the media, that the Polish entrance into the Schengen agreement will be met with fortifying precautions by the Germans due to the disputed German–Polish border history, and on the other hand the relative indifference towards opening the border immediately after the event.

There are strong resemblances between this apparent or relative indifference towards the German–Polish border and what I learnt when I met with the German and Polish high-school pupils, as part of my field research in Görlitz and Zgorzelec, during the spring of 2007 about 6 months ahead of the Polish entrance into the Schengen agreement. These youth did not seem to care much about cross-border activities and the fact that the hyphen German–Polish was often included in posters and other advertisements concerning cross-border activities seemed to contribute to their apparent indifference. This indicates an interesting fact that to live with, across, and at an international border also means forgetting, although momentarily, about it, integrated as it might be into the mundane activities of everyday life (Löfgren, 2008). Yet the question remains, how to account for the indifference and its momentary character and when does the apparent indifference shift into its reverse?

I suggest that this approach of (temporary) indifference towards the border and issues related to the border among local people living there can be understood as effects of the ordering mode of Allocation. Importantly, it is not a constant indifference. Rather, the virtues Allocation performs are the ability of a temporary bracketing of some issues vis-a-vis other issues. The ideal actor (proper person) is focusing on more than one thing at a time and the resources required (means) are the act of shifting issues in and out of focus, thus allocating attention towards issues related to the German–Polish cooperation or border history. Whereas Enterprise and Administration draw on liberal market theory and Weberian sociological insights respectively, the Allocation mode of ordering can be theoretically resourced from multiplicity-oriented Actor Network Theory (Law, 2004; Vikkelsø, 2007; Singleton and Michael, 1993). This theoretical realm seems fruitful in analytically grasping the co-existence of indifference towards and concern about German–Polish issues as patterns of absent and present borders among the German and Polish youth (as in Sandberg, 2009a, 2009b; cf. Law, 2002, 2004).

When Allocation performs the German–Polish border as a (boring) non-issue it stands in sharp contrast to other modes of ordering at stake at the German–Polish border, earlier characterized as working as a cross-border regime, and for whom the border needs to be handled, i.e., addressed, memorized, commemorated, crossed, transgressed, equalized (i.e. economically) on all accounts. This interference between different modes of ordering at the German–Polish border will be the fulcrum for the last and more empirically 'thick' part of my analysis focusing on the mobilization of the youth into cross-border activities.

The Enrolment of the Youth

The young people that took part in my research were, at the time of my fieldwork, 15–17 years of age. Thus born between the years of 1990–1992 they aspire to be a 'post Berlin Wall generation', unburdened by memories of a time when the German–Polish border was removed and replaced, in 1945, and of socialist times of non-communication across the border (Weger, 2005). Being a prominent target for the many integration projects of Görlitz and Zgorzelec (Kurcz, 1999) part of these youth of Görlitz and Zgorzelec attended a bilingual German–Polish educational programme in Görlitz. The focus of the following will be the mobilization attempts performed by the before mentioned cross-border youth coordinator (hereafter the youth coordinator) from Görlitz.

A Process of Responsibilization

The tasks of the youth coordinator include the organization of German–Polish youth projects, fundraising, PR, lobbying and information work. Apart from being sponsored by the municipality of Görlitz the projects are, to a large degree, supported by the EU Inter-regional programmes. According to the youth coordinator the possibilities of creating projects related to German–Polish youth integration in the two cities at the border are very rich. When I visited his office for the first time during fieldwork in 2005, he showed me how his office walls were literally wall-papered with A4 folders packed with brochures, programmes and fundraising possibilities related to German–Polish youth projects. Weekly events include German–Polish bike-trips in the border area, or German–Polish jogging sessions offered ahead of the annual Europe-Marathon, which takes place in Görlitz and Zgorzelec. It is remarkable how the rationale of this seems to be that integration and physical activity go hand in hand; if you are a young citizen living in the region it is possible to bike, run or even paddle yourself into German–Polish integration. However, as the youth-coordinator explained to me, he would prefer that the initiatives came from the young people themselves:

> Almost everything is possible; [there are] interesting, attractive offers – the only precondition is that the young people express their interest themselves, they can

> come to me and then we will work something out. If they want to make a bike-trip, we will make a bike-trip. If they want to canoe, we will go canoeing on the [river] Neiße. … Only, they have to show their own interest … I do not want to create projects and announce them on some poster and then nobody shows up. (Interview 20051026)

Thus, the strategy of mobilizing the youth also encompasses an effort to make the youngsters participate in the planning of activities. The young people's potential thereby becomes part of the integration process, encouraging them to feel individually responsible for it. This process of 'responsibilization', effecting virtues such as 'the ability to accept greater responsibility for oneself and one's actions' (du Gay, 2000, p. 66) contributes to the making of the entrepreneurial actor (cf. Andersen and Sandberg, 2008). By offering the youth involvement in activities – ideally – the individuals themselves get personally empowered. According to the youth coordinator, however, this strategy has not hitherto helped decrease the recruitment problem. Apparently the young people of Görlitz and Zgorzelec do not associate international youth exchange with projects made conjointly with the next-door neighbour. They wish to go to France, Italy or Denmark, as he explained to me. Therefore the youth coordinator uses other strategies as well. Once a month, the youth coordinator produces what he calls his 'Hauptinstrumentarium' [main tool], namely a newsletter that collects and presents all current cross-border youth projects and offers. He had managed to widen the number of subscribers including, amongst others, the municipalities of Görlitz and Zgorzelec, schools, and youth organizations of not only the border area but also broader out into Germany, Poland and the Czech Republic. At the time of the interview, the youth coordinator was also planning to launch new innovations such as a mobile poster pillar that could be 'transported from school yard to school yard, or be exposed in front of the cinemas or the sport clubs of the area' (Interview20051026).

Hence, the youth coordinator can be described as a nodal point in a network of German–Polish activities. He sees all the possibilities and future prospects, it is merely a question of distributing all of these offers from the A4 folders to his target group and enrolling further participants to his network (cf. Callon, 1986). Therefore he seeks back up from other actors, such as mobile poster pillars and the launching of newsletters that can help him distribute his message into still further settings.

The Absent Youth

One illustrative example of the non-appearance of the youth is the activity of the weekly German–Polish bike-trips. With the hope of meeting some of the young Germans and Poles partaking in cross-border activities, I decided to join one of these bike-trips together with my Polish translator Kasia. On the day of the trip, while waiting with our rented bikes, we came to realize that we would be the only participants that day. Later on, when I met with a group of pupils from a Polish school in Zgorzelec, I asked them if they knew about these German–Polish activities and why they rarely chose to partake in them. These conversations took part during a 'walking conversation', the aim of which was to take a walk with the pupils and have them show me their preferred places in Görlitz and Zgorzelec. As the following excerpt shows, it appeared to be foremost the teacher, taking the 'walking conversation' together with us on this particular day, who showed enthusiasm about the cross-border projects and activities:

> Question: 'What kinds of activities interest you; e.g. German–Polish bicycle-trips, the Apollo Theater in Görlitz, Dom Kultury in Zgorzelec, or?'
> Teacher: 'Yes, of course we are interested in activities like that!'
> Question: 'We went on a bicycle-trip last Monday; yet apparently only Kasia [the Polish translator] and me and then the German guy, who arranged the trip, showed up. No one else … I understand that you [the teacher] find these activities attractive, but do you [the pupils] take part in it?'
> Girl 1: 'Well, generally I have no time. When I come back from school, I go for a walk with my dog, or I go to play the guitar, or I go to my sister ... well I have really no time …'
> Girl 2: 'Well, once I took part in a German–Polish drawing contest down by the pedestrian bridge; it was just by accident that I found out about it, and when I saw it, I decided to join! And I really liked it!'
> Question: 'What do you think when you see an arrangement which has "German–Polish" in the title or included in the description of the activity? Would this be more or less interesting for you?'
> Girl 3: 'I think we are indifferent to it ...'
> Girl 1: 'It's nice that they are going to do something together, but it doesn't necessarily have to be us … ' (WC20070523).

This excerpt demonstrates how the addressing of German–Polish relations or cross-border issues does not catch the pupils' immediate attention. This proves to be the case in several examples from my fieldwork material, also among the German pupils. The particular comment stressing that it is fine that 'they' are engaging in common projects 'but it doesn't necessarily have to be us' replaces the focus on the German–Polish integration and the border as issues to be addressed and even transgressed, with other areas of interest. These pupils are aware of the border history and that there, due to this history, might be a need for cross-border activity, yet they do not find it necessary to continually put these issues on the(ir) agenda. Allocation as a mode of ordering is thus not effecting a general disinterest but rather a distribution of when and how it is relevant to occupy oneself with the border. Thus, it can be argued, this is not a matter of pure ignorance, but rather a way of letting the border take the back seat.

Providing an example of when the border turns into a concern, it was evident from my visits to one of the classes of the German–Polish educational programme in Görlitz that bilingual teaching across borders does not automatically dissolve borders and national differences. Pupils coming to school from Polish Zgorzelec literally move across the national border on a daily basis. However, as was evident from my conversations with the pupils of this programme, friendships across borders were rather seldom. Boundaries are crossed and re-activated in a symbolic sense within the classrooms of the German–Polish educational program, when they meet socially (after school and in the evenings) on the public squares of the cities and within the social networks of the pupils. It is in this double sense of border movements – connecting and disconnecting motions as well as emotions (Löfgren, 2002) – that traits of Allocation can be traced.

More Than One Mode of Ordering

So what do these stories and examples tell? On the one hand we see an engaged, yet frustrated youth coordinator who is occupied with the cross-border activities and the participation within these. Prerequisites for this engagement are initiative and self-responsibility as seen in the virtues generated by Enterprise. Yet these virtues butt up against another mode of ordering, that of Allocation, which performs a momentary indifference towards German–Polish integration projects. However, in the practice of the youth coordinator more than one mode of ordering appears to be discernible. For example, the lack of a Polish pendant to the youth coordinator is repeatedly shown as an obstacle for the success of our youth coordinators' goals. As the youth coordinator explained to me, on one of the days that we met in his office, he was close to giving up on getting in contact with 'the Polish side' (Interview 20070626). Not long before our meeting, he invited the Polish mayor to give the introductory speech at the opening of one of the German–Polish youth projects and even though his (Polish) wife translated the letter from German into Polish, he did not receive an answer to the invitation. I asked the youth coordinator why he thought he did not get an answer, and he simply had no clue, which indeed contributed to his frustration (ibid.). Thus two modes of ordering are apparently overlapping each other: On the one hand the German–Polish youth integration would function far better, according to the youth coordinator, if there was a Polish partner of cooperation in Zgorzelec and if all communication were followed up, thus securing due process (as in the logics of Administration). Whilst on the other hand, new methods and ways (such as mobile poster pillars, newsletters, and cross-border bike trips), together with virtues of self-responsibilization, are used in order to gain contact with the youth on both sides of the border, on the basis of virtues and ideas similar to the innovative entrepreneur as the ideal actor (as in Enterprise).

The work of the JCC provides yet another example of how different modes of ordering co-exist. In this case it is interesting to see how the different modes of ordering each generate different versions of cross-border integration in Görlitz and Zgorzelec. As the above example with the Polish mayor suggests, the Enterprise mode of ordering is performing a version of the German–Polish cross-border cooperation that stands in sharp contrast to the logic of Administration. Whereas Enterprise relies on an integration that grows from below, Administration seeks to help with providing the best structures and frames in order to facilitate the integration from above, for example, by establishing symmetry within the functions for cooperative work across the border. This could be a reason why the requests from the German side apparently glanced off on the Polish members of the JCC. These two different versions of German–Polish integration advocating bottom-up and top-down structures respectively are thus practiced in tandem. Hence, through the examples provided by the analysis above, it can be seen that there are more than one mode of ordering performed in the JCC, at the very least the modes of Enterprise and Administration. As conceptual tools that can be imputed to the heterogeneous networks of the social, the modes of ordering explain and unfold how different logics or rationales unfurl, and in this case co-exist next to, as well as butt up against each other. How to grasp this co-existence of simultaneous modes of orderings will be the focus of my concluding section.

Co-Existing Modes of Ordering as Living-in-Tension

The aim of this analysis has been to explore the concept of the modes of ordering by applying it to an analysis of how cross-border integration is practiced at the German–Polish border. By introducing the modes of ordering to border research my aim has been to include a broader (socio-material, decentred and relational) notion of actor than we normally see in practice oriented border studies, and to show that there is more than one mode of ordering at stake in the practicing of German–Polish integration processes in Görlitz and Zgorzelec. The analysis has illustrated this multiplicity of ordering modes with three different examples. In the case of the JCC we saw, that within the work and the decision-making processes at least two modes of ordering were at stake; Enterprise and Administration. In the case of the youth coordinator it was argued that at least two modes of ordering could be imputed to his specific practice; Enterprise and Administration. And finally, the suggested mode of ordering of Allocation showed that there are more than two modes of ordering (Enterprise and Administration respectively) at stake in the German–Polish integration process.

In conclusion, I would now like to discuss how these examples underline how this broader notion of actor as is implied in the modes of ordering can make differences between practices visible in new ways.

Conclusion

To argue that the approach of modes of ordering is particularly relevant for border studies because they distribute agency to a broader repertoire of actors is not to say that the different ways the border is practiced are of a symmetrical character. The increasing strength of Enterprise has, for example, been shown in several studies; although in different empirical settings (cf. du Gay, 2000; Petersen, 2011). Some border practices become more stable than others, depending on how much support they get from other socio-material actors; which yet again can prove problematic for other modes of ordering (cf. Law and Moser, 1999).

Hence, my argument is rather that by paying attention to the multiplicity of the German–Polish border and to the question of how it is ordered differently within practices related to the border, it becomes possible to understand better how the differences work (Dugdale, 1999). Consequently and ideally this ambition thus complies with what Law has described as the approach of offering 'powerful but specific and situated tools for working in and upon particular analytical and political problems in order to know these better and to move them on' (Law, 2009, p. 9). One further interpretation of this formulation by Law is that in understanding such problems it will also become easier to work on them better. By approaching the German–Polish border and the cross-border practices as a multiplicity of modes of ordering we can begin to notice differences in new ways. For example, that Enterprise and Administration both seems to be at stake in the practice of the youth-coordinator illustrates that the modes of ordering are not labelling certain actor-types or personality traits; hence the crucial question is not whether the youth coordinator 'is' the one or the other 'mode of ordering' but how agency is distributed to several actors. As we have seen in the analysis above there are examples of aims at transgressing the German–Polish border, and of ignoring it, as well as trying to make it into an asset. Thus, we see examples of the modes of ordering of Enterprise, Administration and Allocation on both sides of the German–Polish border. This means that the differences at stake in the cross-border cooperation do not always follow the differences usually highlighted, namely the differences (and similarities) associated with the national border.

It is important to stress that the way in which these different orderings interrelate cannot be properly described through metaphors of open conflict. In order to grasp what is going on in the practicing of cross-border integration in Görlitz and Zgorzelec I suggest instead using the metaphors of 'permanent reform' and 'incessant skirmishes' proposed by STS scholars Vicky Singleton and Mike Michael (2003, p. 232). This is due to the mutual intersections between the different modes of ordering that can be interpreted as ongoing interferences between the practicing of different borders rather than as indicators for *either* the success *or* the downfall of the integration across the border in the future and presently. This state of multiplicity has been characterized a 'living-in-tension' by Annemarie Mol (1999, p. 83). With this concept Mol aims at conceptualizing the holding together of differences without asserting a (future) state of integration or assimilation, inferring that the differences are different, yet related (Mol, 1999, p. 77). In some cases they interfere, in others they exist in tandem. I find this concept fruitful for describing a common condition for the modes of ordering described within this analysis because it makes it possible to underline this co-existence of differences as condition where the differences are not aligned.

The various attempts at mobilizing the youth of the two border cities into cross-border activities will probably continue to butt up against other modes of ordering that do not wish to put the border into focus. The differences are there, and they do co-exist, and in some instances they interfere with each other, in other cases they even manage to live parallel lives. How this living-in-tension between the different German–Polish border orderings will unfold in the future is a matter of further empirical investigation. However, leaving room for the Allocation mode of ordering within the cross-border regime, and thus for the possibility of a temporary bracketing of issues related to the German–Polish border and its history, might be a start.

References

Andersen, D. and Sandberg, M. (2008), 'Constructing EuroMan. The Enactment of Regional Identity-Formation in the European Integration Process', *Ethnologia Scandinavica* 38: 39–60.

Becker, F. (2003), 'Ortsbezug und Abwanderung. Kulturanthropologicshe Skizzen zum Transformationsprozess in einer Stadt an der deutsch-polnischen Grenze', in K. Bauer-Volke and I. Dietzsch (eds), *Labor Ostdeutschland. Kulturelle Praxis im gesellschaftlichen Wandel*, Kulturstiftung des Bundes.

Becker, F. (2005), 'Ortsidentitäten im "Europa der Regionen'. Das Beispiel einer schrumpfenden Stadt an der deutsch-polnischen Grenze', in B. Binder, S. Göttsch, W. Kaschuba and K. Vanja (eds), *Ort. Arbeit. Körper. Ethnografie Europäischer Modernen*, Berlin: Waxmann.

Callon, M. (1986), 'Some Elements in the Sociology of Translation: Domestications of the Scallops and the Fishermen of Sct. Brieuc bay', in J. Law (ed.), *Power, Action, Belief: A New Sociology of Knowledge?*, London: Routledge, pp. 196–223.

Callon, M. and Law, J. (1997), 'After the Individual in Society: Lessons on Collectivity from Science, Technology, and Society', *Canadian Journal of Sociology* 22(2): 165–82.

du Gay, P. (2000), 'Entrepreneurial Governance and Public Management: the Anti-Bureaucrats', in J. Clarke et al. (eds), *New Managerialism, New Welfare*, London: Sage.

Dugdale, A. (1999), 'Materiality: Juggling, Sameness and Difference', in J. Law and J. Hassard (eds), *Actor Network Theory and After*, Oxford: Blackwell Publishing.

Europa-Haus (2011), available at http://www.europa-haus-goerlitz.de [accessed 20.10.2011].

European Commission (2007), *Cohesion Policy 2007–13. Commentaries and official texts*, Belgium: The European Communities Publications Office.

Formuszewicz, R. (2005), 'Erfahrungen und Perspektiven in der Zusammenarbeit von Regionen', in A. Wolff-Poweska and D. Bingen 2005 (eds) *Nachbarn auf Distanz. Polen und Deutsche 1998–2004*, Veröffentlichungen des Deutschen Polen-Instituts. Wiesbaden: Harrasowitz Verlag.

Fure, J.S. (2001), ‚*Gutes Zuhause, aber keine Heimat*‘. *War and Post-war Experience, Narrative strategies and Memory of the German Expellees from the Eastern German Provinces after 1945*, PhD thesis: University of Bergen.

Galasińska, A., Rollo, C. and Meinhof, U.H. (2002), ‘Urban Space and the Construction of Identity on the German–Polish Border’, in U.H. Meinhof (ed.) *Living (with) Borders. Identity Discourses on East–West Borders in Europe*, USA: Ashgate Publishing Company.

Großmann, U. (2006), ‘Perspektiven der Zusammenarbeit zwischen Görlitz und Zgorzelec auf dem Weg zur Europastadt’, in *Deutsch-polnische Zusammenarbeit der Kommunen – Beispiel Schlesien*, Görlitz: Bildungswerk für Kommunalpolitik Sachsen e.V.

Görlitz official homepage: http://www.unser-goerlitz.de/baudenkmale [accessed 20.10.2011].

GUS (2007), ‘Concise Statistical Yearbook of Poland’, The Central Statistical Office (GUS), Warsaw, www.stat.gov.pl/gus [accessed: 20.05.2011].

Hannemann, C. (2003), ‘Stadtentwicklung ohne Wirtschaftswachstum: Was verursacht schrumpfende Städte in Ostdeutschland?’, in K. Bauer-Volke and I. Dietsch (eds) *Labor Ostdeutschland. Kulturelle Praxis im gesellschaftlichen Wandel*, Auftrag der Kulturstiftung des Bundes.

Hanneman, C., Kabisch S. and Weiske C. (2000), *Neue Länder – neue Sitten?* Berlin: Schelzky & Jeep.

Heimpold, G. (2001), ‘Regional Policy on the German Border with Poland’, in G. Bucken-Knapp and M. Schack (eds) *Border Matters*, Border Region Studies 2, Institut for Gr[æ]nseregionsforskning.

International Herald Tribune (2007), ‘Quietly, the Polish–German Border dissolves’, by Nicholas Kulish, *International Herald Tribune* 19 December, accessed 20.10.2011

Jajeśniak-Quast, D. and Stokłosa K. (2000), *Geteilte Städte an der Oder und Neisse. Frankfurt (Oder) – Słubice, Guben – Gubin und Görlitz-Zgorzelec 1945–1995*, Frankfurter Studien zur Grenzregion 5, Berlin: Verlag Arno Spitz.

Jensen, T.E. (2001), *Performing Social Work: Competence, Orderings, Spaces and Objects*, PhD thesis, Department for Psychology, University of Copenhagen.

Jensen, T.E. (2004), ‘The Networking Arena’, in T.E. Jensen and A. Westenholz (eds) *Identity in the Age of the New Economy: Life in Temporary and Scattered Work Practices*, Cheltenham: Edward Elgar Publishing, pp. 55–78.

Kramsch, O. and Hooper B. (eds) (2004), *Cross-border governance in the European Union*, London: Routledge.

Kurcz, Z. (1999), ‘Polnische Grenzregionen im Vergleich’, in *WeltTrends* No. 22: pp. 63–74.

Latour, B. (1996), ‘Om aktør-netværksteori. Nogle få afklaringer og mere end nogle få forviklinger’, *Philosophia. Tidsskrift for filosofi* 25: 3–4.

Law, J. (1994), *Organizing Modernity*, Oxford: Blackwell.

Law, J. (2000), ‘On the Subject of the Object: Narrative, Technology, and Interpellation’, *Configurations* 8(1): 1–29.

Law, J. (2002), 'On Hidden Heterogeneities: Complexities, Formalism, and Aircraft Design', in J. Law and A. Mol (eds), *Complexities: Social Studies of Knowledge Practices*, Durham, NC: Duke University Press.

Law, J. (2004), *After Method: Mess in Social Science Research*, London/New York: Routledge.

Law, J. (2009), 'The Greer-Bush Test: on Politics in STS', version of 23rd December 2009, available at http://www.heterogeneities.net/publications/Law2009TheGreer-BushTest.pdf [accessed 20.10.2011].

Law, J. and Moser I. (1999), 'Managing, Subjectivities and Desire', Centre for Science Studies, Lancaster University, Lancaster LA1 4YN, U.K. http://www.lancs.ac.uk/fass/sociology/papers/law-moser-managing-subjectivities-desires-pdf [accessed 20.10.2011].

Linde-Laursen, A. (2010), *Bordering. Identity Processes between the National and Personal*, Aldershot: Ashgate.

Löfgren, O. (2002), 'The Nationalization of Anxiety: A History of Border Crossing', in U. Hedetoft and M. Hjort (eds) *The Postnational Self. Belonging and Identity*, Public Worlds, vol. 10, Minneapolis: University of Minnesota Press.

Löfgren, O. (2008), 'Regionauts: The Transformation of Cross-Border regions in Scandinavia', *European Urban and Regional Studies* 15(3): 195–209.

Mol, A. (1999), 'Ontological Politics: A Word and some Questions', in J. Law and J. Hassard (eds), *Actor Network Theory and After*, Oxford: Blackwell Publishing.

Mol, A. (2002), *The Body Multiple: Ontology in Medical Practice*, Durham, NC: Duke University Press.

Mol, A. (2008), *The Logic of Care, Health and the Problem of Patient Choice*, Abingdon, Oxon: Routledge.

Moser, I. (2005), 'On Becoming Disabled and Articulating Alternatives: The Multiple Modes of Ordering Disability and their Interferences', *Cultural Studies* 19(6): 667–700.

Petersen, M. (2011), *'Good' Outcomes. Handling Multiplicity in Government Communication*, PhD thesis, Department of Intercultural Communication and Management, Copenhagen Business School.

Regionaler Planungsverband Oberlausitz-Niederschlesien (2011), http://www.rpv-oberlausitz-niederschlesien.de/planungsregion.html [accessed 20.10.2011].

Rumford, C. (2006), 'Theorizing Borders', *European Journal of Social Theory* 9(2): 155–69.

Rumford, C. (2008), 'Introduction: Citizens and Borderwork in Europe', *Space and Polity* 12(1): 1–12.

Sandberg, M. (2009a), *The Present/Absent Border. Europeanization Processes in the Twin Cities of Görlitz and Zgorzelec at the German-Polish border*, PhD thesis, University of Copenhagen (in Danish, can be accessed at www.dasts.dk).

Sandberg, M. (2009b), 'Performing the Border. Cartographic Enactments of the German–Polish Border among German and Polish High-School Pupils', *Anthropological Journal of European Cultures* 18: 107–28.

Schultz, H. (2001) (ed.), *Grenzen im Ostblock und ihre [Ü]berwindung*, Frankfurter Studien zur Grenzregion 6. Berlin: Arno Spitz Verlag.
Schultz, H. (2002): 'Twin towns on the Border as Laboratories of European Integration'. Frankfurter Institut für Transformationsstudien 4/2002. http://www.isigmagazine.isig.it/lib/files/isig_mag_bollettino_it_2270_pdf_.pdf [accessed 03.10.2011].
Singleton, V. and Michael, M. (1993), 'Actor-Networks and Ambivalence: General Practitioners in the UK Cervical Screening Programme', *Social Studies of Science* 23(2): 227–64.
Stadtverwaltung Görlitz (2006/2007), *Statistiches Jahrbuch 2006/2007*, Kreisfreie Stadt Görlitz, der Oberbürgermeister, kommunale Statistikstelle, Stadtverwaltung Görlitz.
Van Houtum, H. and Naerrsen, T.V. (2002), 'Bordering, Ordering and Othering', *Tijdschrift voor economische en sociale geografie* 93(2): 125–36.
Vikkelsø, S. (2007), 'Description as Intervention: Engagement and Resistance in Actor-Network Analyses', *Science as Culture* 16(3): 297–309.
Weger, T. (2005), 'Görlitz und Zgorzelec nach 1945', in E. Opilowska and S. Pfeiffer (eds), *Görlitz und Zgorzelec – zwei Seiten einer Stadt*, Sächsische Zeitung, Hrsg. Michael Wieler, Dresden.

Interviews and Walking Conversations (WC)

Interview20051026: Coordinator of German–Polish youth activities. Interview made in Görlitz 2005 and translated by Marie Sandberg.
Interview20070525: Polish mayor of Zgorzelec. Interview made by Marie Sandberg in Zgorzelec 2007 and translated by Katarzyna Wala.
Interview20070626: Coordinator of German–Polish youth activities. Interview made in Görlitz 2005 and translated by Marie Sandberg.
WC20070523, made in Zgorzelec by Marie Sandberg 2007 and translated by Katarzyna Wala.

Chapter 7

The Multiple Politics of Borders: Images of the Slovenian–Croatian Border on Istria from the Perspective of an Ethnographer on the Move

Dorte Jagetić Andersen

Introduction

When Yugoslavia fell apart for the second time in its history, the geographical area became subject to a tremendous amount of change, including the drawing of new and more borders between independent states (Lampe, 2000). The first border to be recognized by the international community, today separates the republic of Slovenia from that of Croatia. As a consequence of the history of the region, the fairly quiet Slovene secession from Yugoslavia as well as the early international recognition of Slovenia and Croatia's right to declare the states independent, the location and existence of the Slovenian–Croatian border is rarely regarded as geopolitically arbitrary. The institution of this border, which is for the time being also an external EU and a Schengen border, nevertheless changed life for people living at the border. Even when the physical existence of the border is not questioned, it is and remains a geopolitical construction intruding in everyday life and interaction in the regions located at it (cf. Barbic, 2004; Pavlakovich-Kochi and Stiperski, 2004).

This chapter shed light on a set of border practices located in intersections and tensions between geopolitical decisions and the everyday life of people living on the Italian, Slovenian and Croatian peninsula, Istria. Due to the history of this peninsula, much of the reasoning about borders in the area reach far deeper than the geopolitical decisions dividing the peninsula into three different states. Still – and perhaps therefore – borders and geopolitics has an impact on life in the region. The peninsula is characterized by the presence of a multitude of nationalities, not just as an effect of its threefold division but also because tourism is the most important economic foundation for people here. Everyday life on Istria is lived without apparent confrontations between national differences, yet identity-discourse – supposedly confined to a different space than that of everyday life – stabilizes the images of differences between nationalities including the image of a Croatian–Slovenian border, which separates two nations consisting of different people with different outlooks on life.

Taken together the border practices illustrate how everyday life at this border is far removed from geopolitics, at the same time as they illustrate conflicts generated by as well as what is at stake for people when geopolitical decisions are made. Since the 1990s, public policy approaches in border studies have, in line with EU visions, concentrated their investigations of border regions on cross-border partnerships, cooperation and networks of actors, which are deliberately engaged in dismantling borders (Perkmann, 2000; Scott, 2006, 2010; Boedeltje et al., 2006). These networks consist of what we would usually term political agents, such as local politicians and professional consultants who supposedly move the developments of public policies in border regions. However, if we want to understand the politics of borders we must add to these considerations by investigating politics as ways in which borders are lived, crossed and narrated in everyday life (cf. Pavlakovich-Kochi et al., 2004). We need a way of studying regulatory practices, which takes into consideration that politics is not simply conducted in accordance with EU policies and governmental institutions, implemented by bureaucrats and other professionals. The picture is much more blurred; borders act symbolically to constitute and divide national identities, but local sentiments may also be deeply interconnected with regional nationalisms, and inevitably used to further the interests of border regions; or borders might be understood as irrelevant for 'real life', and thus as artificial constructions intruding with what we believe we are (cf. Andersen, 2011). Borders are enacted and performed at various levels, including geopolitics, public and political discourse, historical narrative and everyday life practices and public policy approaches thus only captures a fragment of what is at stake at and, not the least, with the borders of Europe.

Great decisions might not be taken in Café Europa[1] but rest assured that everybody here has an opinion and over time such opinions might come to count more than the actions of the above mentioned political agents. Hence, there is a need to tell stories from the ground, so to speak, and to emphasize the ability people have to participate in the making of borders as well as the regulations and empowerment that can result from border activities (cf. Rumford 2006, 2008). Borders are also practices that situate and constitute in the everyday life performance of them (Galasinski and Meinhof, 2002; Meinhof and Galasinski, 2002; Meinhof, 2002; Schultz et al., 2002; Sandberg, 2009) and in this dimension we inevitably find things that public policy approaches do not discover.

By illuminating the particular but also very different meanings people ascribe to borders dependent on their interactions with them, the chapter shows how

1 Café Europa is used as synonym for a semi-public, semi-private sphere in which citizens and ordinary people come together to discuss politics. Café Europa has a long history, reaching back to 18th century France where the bourgeoisie raised their voices and refined their opinions of societal organization in the cafés of Paris and other French cities. Thus, every European city has its Café Europa – its meeting place for its cit(y)zens.

culture and politics interlink in the concepts we live by and not just in the political act itself. These intersections between culture and politics as well as established institutions and lived life, are important mechanisms to capture if we want to understand the significance of borders in a globalized context (Pavlakovich-Kochi et al., 2004). However, as their significance cannot be captured if we do not possess tools to investigate borders beyond static notions of politics, territory and identity, I will first clarify the approach taken in this particular investigation before I move on to the empirical analysis.

Ethnography on the Move

Border studies depend to a large extent still on un-clarified presuppositions about the nature of borders when coding and making summations in their empirical studies (Pasii, 2005, p. 28). This counts foremost for links between borders, state territory, nations and national identity. For instance, due to the history of the European states as well as their relation to national identity-formation, border studies have a tendency to focus on this particular identity construction and thus dissociate from other forms. National identity and state histories may explain issues in retrospect but their ability to illuminate what takes place at the moment might not be as significant as we want to believe. The contingency of real time events may in other words overwhelm the patterns created by events passed, or as Garfinkel reminds us: 'Predictability arises not from "big" causal factors that precede (and might predict) situations of work but from everyday, mundane (and yet remarkable) accomplishment of those people and things involved' (Garfinkel, 1967).

Borders are in other words not just enacted in wars between states and in public policy-making. Rather, the multiplicity of acts performed in intersections between culture and politics continuously disseminate, dissolve, disperse and construct borders between us. Borders exclude and include, they are functional and symbolic at once, they help us construct national, regional, local and European identities, they generate conflicts between people at various levels of interaction – and sometimes they even generate conflicts within people. As Löfgren notes: 'Among the different ways of organizing experiences, identities and communities, the pedagogy of space is very striking; borders as markers of territories, transitions and passages.' (Löfgren, 1999, p. 6) Images of borders thereby provide hard evidence for the fact that culture and politics connect in the concepts we use to articulate who we are, concepts we also act upon.

At the same time and perhaps paradoxically we cannot simply neglect images of borders, which depend on state divisions and form in sentiments of national belonging reflecting subjective associations to national history. As Anssi Paasi puts it: 'Boundaries and their locations are often crucial elements in representations and narratives regarding the past successes and defeats of states and nations, on account of the fact that during the 19th and 20th centuries boundaries and territories became political symbols over which "nations went to war and for which citizens fought and died" (Sahlins, 1989)' (Paasi, 2005, p. 20). This is one side to the story; the ways in which state borders have helped construct narratives of national belonging by physically demarcating the territories in which these identities were confined. Another side to this story is how the numbers and figures which are used in quantitative research today are most often confined to the exact same territories: 'Most bodies of comparative data on human societies are gathered from statistics created by nation-states and are interpreted at the state level, often using maps depicting boundaries, which effectively reifies the existence of bounded territorial space' (ibid, p. 21). It is therefore also important that border studies do not dissociate from a world divided into state territories but take seriously the significance of all images of borders, how they are lived and performed as well as how they affect the politics of borders.

Hence, to investigate borders, method are needed, which can illuminate how an abstract concept – the border – is given meaning in a multiplicity of enactments, performed in local context as well as how such enactment and performances continuously transforms the abstract concept by relating it to other concepts – identity, nationality, statehood and personhood – and enacting it in real time. Moreover, in the enactment of borders their Janus Faced significance has to be recognized. On the one hand, borders connote what keeps the other out, a boundary between my world and that of others; on the other hand, it opens up towards the sensation of a world without limits and thus without boundaries constituting identity. The Janus faced significance of borders inevitably links them to the political act; images of borders always occur in a choice between a world of total hospitality and a world of regulatory practices, of including and excluding:

> A tension thus lies at the heart of performative border spacing, which reveal on the one hand practices of control, the production of inside and outside distinctions, the ongoing carving up of domains of knowledge and purified 'dreamlands' of id/entity (van Houtum, 2002); and on the other an escape into radical openness, into teeming border-crossing inventiveness (Kramsch, 2002). (Van Houtum et al., 2005, p. 3)

Inherently, the border is a contradictory socio-material ordering of space with political connotation and somehow always situated geographically as well as in time. Only choices made by people, including the processes in which meaning is ascribed to this contradictory concept may undo the contradiction, or at least hide its significance to the spectator. Hence, the aim of border studies must be to decode borders, not as static entities bound up with specific self-evident concepts but as mobile, travelling borders enacted in space and time with all the contradictions and conflicts this may entail. With Hinchliffe's words, we have to focus on: 'The assembling practices that associate diverse collections of people and things, and on the assembling of practices that these people and things then allow' (Hinchliffe, 1999).

Hence, to grasp the politics of borders I employ what I call 'an ethnography on the move', that is, a method of participant observation moving along with its object of study – in this case the border and, more specifically, images of borders deriving from the practice of them, how they are enacted, performed, and lived. The ethnography moves, foremost, in a physical sense in that it moves along with the object-subject of research whatever its status might be and wherever it might go. The border may be encountered by physically existing Slovenian-Croatian checkpoints but if this border moves to Stockholm, Brussels or to the internet, then we must follow it to understand the practices and narratives constituting it. The researcher follows the border in the act, so to speak, thus clarifying how borders are enacted by individuals through talking, walking, writing and a variety of other practices (cf. Sandberg, 2009; Laurier, 2011; Zaiotti, 2010). When the border is 'on the move', so to speak, there may be many borders at stake at once in the same real time. This is to say that borders are practiced differently by different actors and this multiplicity may conflict when practices link in the same space of ordering (Byrska, 2012; Andersen et al., 2012; Andersen, 2011) – hence, the implicit politics of borders capturing their contradictory significance.

Ethnography on the move locates in the practice turn referred to in the introduction to this volume and is inspired by Actor Network Theory's quest to 'follow the actor' (Latour, 1986, 2008; Sandberg, 2009). However, instead of tracing those network relations, which have come to constitute black-boxes (Latour, 2008), it focuses on multiplicity and enactment (Mol, 2002; Sandberg, 2009, herein) and it recognizes the importance of hierarchy and institutions, as in traditions, habits, social patterns, history and social structures, which guide practice. Consequently, the movement also has to take place between structural and institutional practice – between for instance the EU, the state, the nation, the region, the locality, civil society and the family. All of these inform border practice (Donnan and Wilson, 1999; Wilson, herein, Andersen et al., herein) foremost in the sense of a dispersion of discourses relating to stratifications of culture, identity, community and belonging. Hence, the movement is also one between the insights of different academic disciplines and stylistic genres. Anthropology, history, sociology, politics, literature and film are all part of the stories told and rather than analytically separating these in order to grasp 'fact' and 'detail', it is necessary to approach lived life as infinite webs of relations in which we, as researchers, must situate ourselves and navigate. Ethnography on the move takes this multiplicity into account by illuminating how borders locate in the imaginings and practices of the 'more than one, less than many'.

Moving beyond presuppositions is of course not an inevitable result of doing ethnography. Often so called ethnographic studies do not pay sufficient attention to their most fundamental method, namely participant observation, and instead they replace it with the more systematic interview method (cf. Wilson, herein). As Laurier puts it in his study of mobile workers: 'All too often socio-spatial investigations that claim to be "ethnographic" proceed through an overformalized, scientized process of interviewing, transcribing and coding while setting aside the "under-built" method of participant-observation' (2011, p. 3). By contrast, ethnography on the move emphasizes participant observation as the basis of ethnographic work, even when oral and written communication both are part of the tool box of the ethnographer: 'Ethnography's credo is to ask people what, why and how they do what they do and to compare that to what they actually do' (ibid, p. 3). Of course, the informants have voices but they are not considered to use categorizations in such a way that it is actually possible to code the conversations with them. Hence, ethnography on the move takes seriously the hermeneutic tradition within which ethnographic studies locate, recognizing how researchers construct the narratives told, thus positioning themselves as spectators and participants at once (Clifford and Marcus, 1986). The researcher therefore needs to recognize those on behalf of whom he speaks and reflect this insight in his writings. Our limitations as researchers are thus not to be understood as an invitation to do fiction but rather as a realization that we, as researchers, always speak on behalf of those we understand to be important.

In this article I speak on behalf of Café Europa or, more precisely, local people living in as well as people located at, visiting or otherwise having contact with the Croatian–Slovenian border at Istria. However, the voices of those on behalf of whom I speak are not the ones telling the actual narratives. Rather, the narrative structure of the reporting is done by me thus also bringing together voices, which have not actually met in real time (at least not to my knowledge). My aim has been to depict a socio-spatial setting in which various actors create multiple, yet related borders when they use them, control them, cross them, think of them or in other ways relate to them practically. These borders – be they inclusive or exclusive, open or closed, Janus faced and contradictory, or simply irrelevant – articulate both the political core to borders as well as conflicts generated with their constitution. There are of course more narratives to be written but the few told here already illustrate how deeply connected the multiple politics of borders is with issues commonly associated with cultural practice and understood as separate from the realm of politics proper.

The data was gathered and narrated by me during fieldwork in Northern Croatia in periods from October 2009 to August 2011 where I lived in Zagreb and went to Istria on a regular basis. It was gathered in accordance with an ethnographic method moving along with local narratives articulating images of the border, also when these turned into public discourse or political disputes about the border. Some voices are partly known to the Slovenian and Croatian public – one is for instance an official dispute – whereas others come from individuals who practice borders because of their occupation, the location of their home or simply just their name. None of my narratives have prevalence over others, even when some of them might be more known, popular or disputed than others. What does have prevalence, however, is the understanding of the border as a multiple, political and contested space of ordering – reflecting the multiple politics of borders.

The Border Space

The territorial state border between Slovenia and Croatia is of course very present at the border control posts on this newly erected dividing line. Arriving on Istria and not knowing where to look for borders, I therefore went to one of the control posts, Plovanija, the one furthest to the west of Istria. Not only the Croatian border authorities, allowing me to be physically present at the control post, but also the different license plates of cars passing through immediately reminded me of a world divided into states and state territories. However, when asking people crossing the border about their experiences with it most of them did not associate it to state divisions, let alone differences between nationalities. Rather, the border control post infers 'control', 'queues', 'hassle' and it is 'time consuming'. The control post constitutes a space separate from normality, a space of exception, which is not immediately connected with what goes on within the state territories it is meant to control (see also Gawlewicz and Yndigegn, Chapter 9 herein). Moreover, it is not a pleasant space but a space people are eager to pass through quickly in order for them to return to normality; the stiff look on people's faces when going from one side of the control post to the other indicates this and cannot be combated by the fact that they are protected by the safety of their cars from whatever violence the border might impose.

A more specific question I asked people crossing the border was what their associations and experiences were with European borders. What happened for many of them at first was that they could not find any answers and only when I emphasized the word 'European' a few times they said: 'But they do not exist'. Being in the border space, the border has no relevance let alone existence if it is not indicated by a checkpoint, confronting you with regulatory practices. The border space is thus constituted by a unique checkpoint culture (see also Byrska, Chapter 5 herein).

When asked about their opinions on the existence of the border control post at Plovanija and whether it was beneficial that it is to be removed with Croatia's accession into the European Union in 2013, the opinions were divided. Many people crossing saw the control as a necessary evil, being somewhat indifference to it; they would come whether it was there or not. By contrast, a group of young Germans said they would definitely consider it beneficial if there was no border – but they had also been held up for half an hour at Plovanija while being checked for drug possession. A man from New Zeeland began complaining about European borders being much too closed as it were, when taking into consideration that they were supposed to be open, – whereas a Danish couple thought it was very reassuring to have a control post at the border and that they would feel safer if more control was reintroduced in Europe. A British truck driver on holiday with his family saw both pros and cons of the control. Once he was stopped at a border with refugees in his truck, something, which he was himself not aware of and would not have been discovered had he not been held up. At the same time, it had made life much easier for truck drivers in Europe now that the physical control is limited and most control happens electronically. Seen from the control post, the border constitutes an isolated space, whose existence is highly contested, at the same time as it somehow orders people to belong to state territories.

The Territorial Border

Despite its history, political and public discourse, in Slovenia and Croatia, rarely asks the question if the existence of a border between the two states is actually more of a coincidence than a historical fact. Hence, the border functions to separate national populations, who are different. Paradoxically perhaps, this also counts for many locals on Istria. The stereotypes local people invent about different nationalities visiting the peninsula know no limits; Italians are loud; Germans are correct; Austrians are peasants; etc. About Slovenians the saying goes that they invade the peninsula in their big caravans, taking their food with them and buying nothing from the locals. Moreover, wherever they go they leave behind all their garbage, be it on the beach or on the camping sites. The discourse constructing national differences between Slovenians and Croats is in other words extremely vital and in recent years, the discourse gained substance due to conflicts between the two countries over accession to land in the Piran Bay (see below). It seems natural today, independent on whether or not one felt part of a larger Yugoslav union, to discursively construct the image of two populations, which are fundamentally different, both in future aspirations and because of their geopolitical situation in the past. Here the physical existence of the border plays a vital role.

A story that caught my attention while I was living in Croatia reflects these nation building tendencies. It is a story about a middle-aged man, Joska Joras and his house, captured in a dispute with the Croatian and the Slovenian state. Joska Joras is a Slovene, who lives in Croatia. However, Joras is famous in both Croatia and Slovenia because he raised the question of whether he does actually live in Croatia and not in Slovenia. He provides maps, local population registers and his version of local history to back up his claim that his house is in Slovenian territory and he refuses to follow Croatian jurisdiction (Ljubljana Life.com, 2008). Joras' house is located on the south side of the river Dragonja and the stretch of land is considered Croatian territory. Ownership of the strip of land is officially unresolved, however, even when Slovenia conceded the house to Croatia because of the dispute with Joras. Joras' reason for wanting to live in Slovenia is personal:

> I feel a duty to respect the heritage of my family,' he says, revealing that the house where he has lived for 40 years belonged to his mother who was kept in a concentration camp for three years during Nazi occupation of Yugoslavia. 'I feel a duty to maintain the integrity of the country for which she was sacrificed. (Ljubljana Life.com, 2008)

Hence, Joras flies a Slovenian flag on the side of his house, which the Croatians say is done as a provocation and he has a sign facing the Croatian border that reads in large black letters: 'This is Slovenia'. Moreover, Joras does not pass through the Croatian border post but uses a grave track between the two border posts to get to his house.

Generally speaking, Croats and Slovenians do not think of Joras as anything more than a mad old man with nationalist aspirations. However, Joras is not alone. He is backed up by the Slovenian People's Party whose leader once came to visit him:

> One of the funniest little side shows to this whole situation was played out a couple of years ago when the president of the Slovenian People's Party, Janez Podobnik, decided to pay Mr. Joras a visit without observing Croatian border formalities. It all ended in a big scuffle involving the Croatian police, and with Podobnik's brother actually punching one of Croatia's finest on the nose. The group of Slovenians was arrested and spent some time in a Croatian jail, causing a big international incident. (Ibid.)

And when I spend time at the border control post, Plovanija, talking with people crossing the border as well as the police and customs controlling it, I could not help but to notice a house facing the control post flying a Slovenian flag and upon which it said with big letters: 'Tudi Tukaj je Slovenija' (again, Slovenia). The house is located in what is officially Croatian territory but only a few yards away from the no-man's-land that divide Croatia and Slovenia at Plovanija.

The (In)determined Border

When public opinion criticizes Joras, it is not the physical existence of the control posts on the Slovenian–Croatian border, which is questioned. Rather, the critique concerns Joras' determination to prove that his particular border is misplaced. However, the physical existence of the border is actually not as natural as many people including tourists crossing the control posts would like to believe. Foremost, the entire border was never formally drawn, neither in the context of Yugoslavia nor in negotiations between the two states. This is the case with the important sea territory of the republics: 'In the second Yugoslavia, there was no delimitation of the sea between the republics, which is why also the sea border in Istria between the former Republic of Slovenia and the Republic of Croatia was not determined.' (Krnel-Umek, 2005, p. 34) It is only parts of the border that have been renegotiated after 1991, something, which was the result of political disputes occurring during the 1990s and 2000s. In one of these disputes, Slovenia demanded of Croatia to give up parts of its sea and coastal territory if the country was to be accepted into the European Union. Slovenia forced Croatian politicians to enter renegotiation of the border along the coastline, because Slovenia did not

Figure 7.1 'Again, Slovenia'

have direct access to international waters and has to pass through either Italian or Croatian sea if they wanted access. Slovenia, with its very limited coast line, first claimed its right to more sea territory and parts of the coastal line in Istria. Later the claims were moderated to only concern sea territory (BBC News, 2009). Today, the border dispute is being settled by an independent commission joined by the European Union proposing to include parts of Croatian sea territory into Slovenia in order for Slovenia to gain direct access to international waters and without taking over any of the Croatian coastal territory.

Nevertheless, the questions remain: Is it the Mirna that divides Croatia and Slovenia? Is it the Dragonja and where exactly by the Dragonja? Or is it in the middle of the Piran bay that the line ought to be drawn? Historians provide documentation and argue for where and why the border should be drawn. The histories are based on different kinds of historical evidence, mainly focusing on ethnic divides on Istria or, from the Slovenian side, the fact that a new negotiation of the border is needed because of new state divisions in the area. The stories also differ tremendously, as one might anticipate, depending on whether they are written by Slovenian or Croatian historians (Krnel-Umek, 2005). Such histories not only illustrate how valid historical sources can be used in different ways considering the interests involved; they also exemplify how contested state borders are.

In my conversations with the border police, whether it was those controlling the land or those controlling the sea, the general opinion was that they wanted the border dispute to settle. Obviously it makes their work harder when they have to control borders, which are not yet determined. But as one of them said to me: 'Borders are all in the head, and when we realize that, we can also diminish their importance.' For him, working with borders in his everyday life, borders are not natural dividing lines between distinct nations and populations who have an essential and stable national identity rooted in historical soil. Rather, borders are constructs of geopolitical decisions, and not the least, of history, un-folding and de-folding in discourses that might not have much to do with so called reality but nevertheless secure their existence. In the Slovenian–Croatian case, the border is very visible today because new state-territories, including new divisions between political systems and locations of decision-making powers, have to sediment. Despite the fact that many want it to be and others understand it as such, the function of this particular border is thus not very solid and, practically, the border changes almost on daily basis in constant attempts to solidify it.

The Ideological Border

When visiting Istria and talking to the local populations, you will not be confronted by many people like Joska Joras, but by images of the border influenced strongly by the sense of familiarity. The populations living on either side of the border have had a peaceful relation for years and in the everyday practices of people 'the other' – as in the Slovenian other or Croatian other – does not often articulate unfamiliarity. In a deliberate (or perhaps desperate) attempt at the beginning of my visits to Istria to make Croats articulate some form of unfamiliarity with Slovenes so as for the border to appear, I began referring to the conflict over access to international waters (see above). However, despite all the efforts I made, even here the Slovene stayed familiar to the Croats. Rather than distinguishing between themselves and the Slovene population it was politicians and a few more eccentric Slovenes who stuck out as being different.

One category of people who may be considered as immediately affected by the border dispute, are fishermen on either side of the Piran Bay. It is the impression of the locals on the Croatian side of the border that these fishermen have much to say about the dispute as the settlement of it would in some way interfere with their fishing rights. I therefore went to talk with them about their experience of the dispute and about any problems they might have because of it. What I found, however, was that they did not experience any problems. One fisherman told me that there are about 10 Slovenian fishermen using the Piran Bay and they come and go as they like. This is not considered a problem for the Croatian fishermen using the bay and the Slovenians have no problems with them. In his eyes, the

Figure 7.2 Fishing boats in the Piran Bay

border dispute was all about politics and for him completely unconnected with their everyday lives and work as fishermen.

When I mentioned the border dispute to a restaurant owner in Umag located in Croatia, approximately three kilometres from the Piran Bay, he began complaining to me about a residential area occupied mainly by rich Slovenians but on the Croatian side of the bay. The area is surrounded by a fence, making it impossible for people who are not living in the area to access it with cars (you need an entrance key to open the fence) and hence they cannot access the beaches in this part of the Piran Bay. According to the restaurant owner, the problem is not only that Slovenians occupy Croatian land (his words) and exclude local residents from access to the beach, which is public area in Croatia unless you pay a leasing fee to the local municipality. It is also that none of the two governments interferes with the Slovenians and other rich people doing this. People living on Istria do not feel they have any support from either of the two governments, nor from the EU.

In general, however, there are no problems between the Slovenians and the Croatians in the area: 'I do not know what they say on television but we have no problems with the Slovenians. Most Slovenians just come here to get drunk because it is cheaper than in Slovenia.' The restaurant owner continues to recommend that Croatia joins the EU, as he says: 'They cannot take the land if both Italy and Croatia are part of the EU.' In line with this argument, a restaurant waiter in Umag told me:

> But I do not know what they want ... Should Croatia send documents with no Croatian-Slovenian borders drawn at all? Or do they want us to put the borderlines as they want it? If the latter is the case, then it is clear blackmail. The problem is also that I can't see those documents anywhere on the internet. I tried to look for it – no trace of them – neither at the Croatian government site about negotiating process, nor at the EU site.

People living on Istria were afraid of loss of land while the border dispute was on, but what they consider most problematic is how politicians contradict themselves and invent conflicts where they do not exists. Slovenian politicians for instance want to be part of a Europe without frontiers but still refuse to dissolve frontiers when their interests are at stake – leaving the Istrian populations in the middle of a border dispute, which is not theirs. This distance to political decisions is part of a historical memory in an area where ideology has prevailed and most often without any apparent relation to lived life. For many people living in Istria, it seems that issues of importance are always settled somewhere else and by someone else. They do not make problems out of their differences; it is others who do that. And the other is ideology and the political elite, including rich Slovenians taking advantage of politicians and lawgivers incapable of carrying out their jobs. The fact that this border will probably change significantly in a few years when it becomes an internal EU and Schengen border simply proves the fact that people living on Istria are captured by political discourses in relation to which they have very little say. The border is thus an ideological construction.

The Absent Border

In her book, *Cafe Europa*, Slavenka Drakulic describes an incident in 1994 where a Croat reporter interviewed inhabitants of three Croatian border villages on Istria, which had been annexed overnight by Slovenes:

> [He] was faced with a strange reaction. To the Croat reporter, these people said that they were Croats, to Italian reporters that they were of course Italians and the Slovene reporters were told they were Slovenes. To the reporters it was impossible for one person to be a Croat, an Italian and a Slovene all at the same time. (Drakulic 1996, p. 163)

For the villagers, the reporter misunderstood the question of national identity because he took it as an either-or question. Many people living on Istria consider their identity as more than having one nationality and many of them are unable to choose one nationality as their identity:

> Living in the border region, they understand better than anyone else that we all have mixed blood to a greater or lesser extent. They have also suffered from

> nationalism, and in its worst form – ethnic cleansing – enough to have grown tired of it. (Ibid, p. 164)

Hence, the so-called 'Istrians' imagine their regional belonging as their identity:

> When you descend from Crni Kal to Socerga, from the mountains to the valley, you will get a sense of what this identity is all about. Here the landscape does not change, people understand each other's languages and eat the same food, and yet everything is overshadowed by this newly erected monster – the iron construction called the state border that divides people. (Ibid, p. 161)

For many locals on Istria, the presence of a closed territorial border between the peninsula's Slovenian and Croatian parts therefore appears highly arbitrary. Speaking with locals you get the impression that they generally want to be left in peace from geopolitical questions dividing people into states. The impression left on people by political history was not positive and they were not spared by it, having experienced violent nationalisms and genocide. Also in recent years, geopolitical changes have altered everyday life in the region, and especially tourism, the economical foundation of most people on the peninsula, has been affected by the changes. Queues at the border control post are now relatively long during the tourist season. When the area was part of Yugoslavia the queue of tourists was more limited and it was formed on the Italian, not the Croatian–Slovenian border. In the Yugoslav parts of the peninsula, the population lived a life where the checkpoint only appeared when they went shopping in Trieste.

European integration might of course provide a solution to some of the problems experienced by people. Expressed with words from an Istrian inhabitant (quoted in Slavenka Drakulic' *Cafe Europa*):

> Imagine, he says to me, one day perhaps only a few years from now both Slovenia and Croatia will be members of the European Union. All these papers and tensions, all our fears and insecurity will suddenly be obsolete. And no one will force us to identify with just one part of what we experience as our identity. I dream about that day, when nobody will hate me because of the food I prefer, my memory, or the language I speak. (Ibid, p. 169)

European integration has had and will have an impact on life on the peninsula, yet the rationales guiding life at Istria reaches far deeper. Euro-sceptics on the peninsula would even claim that Europe, in the name of European Integration, created most of the problems that it now presents itself as a solution to.

The border, which makes one part of Istria external to Europe proper and the other internal, without doubt appear in the form of a geopolitical construct intruding in and disturbing everyday life and interaction in the border region. Yet life seems to go on somewhere beyond the presence of borders, as it has been going on for generations despite the turbulent history of the peninsula. In the everyday life practices of Istrians, these borders are part of life, as the Istrian regional identity is both shaped by and lived beyond territorial divisions of states and empires. Doubtlessly, the border interferes with their lifestyle; it is an obstacle making border practices less easy to perform. However, Istrians seem to somehow institute a distinction between those border practices, which take place in their everyday life and then the border practices of regulation and control instituted by geopolitical decision making and experienced as located in a space different from everyday life. In this sense the border becomes absent as regulatory practice.

The Border as Asset to Life

There might be an (more or less) official border dividing Slovenia and Croatia on Istria, but the 'other' is not really an 'other' here. Rather, they are part of us, maybe with a few differences but they basically act like us and we recognize ourselves in them. People understand each other on a deeper level and nationalities can even be multiple as is the case with Karlo-Carletto who is Croat in the morning, before he starts drinking, and Italian in the late afternoon and evening, after a few drinks (Drakulic, 1996). The regional identity is also materially present on the peninsula. It characterizes the architecture, which is strongly influenced by the region's Venetian and Roman heritage, the landscape of the region, which does not change whether you are in Slovenia, Croatia or Italy, in the customs of people living here, in their many languages and in the food eaten.

On the one hand, this presence and parallel existence of differences makes the borders invisible, as they are indeed somewhat irrelevant to everyday life. On the other hand, however, borders are an asset to life expressed in the sense of belonging in a context of differences. The Slovenian and Croatian border police for instance have a long tradition for collaboration, something, which makes their work much easier on either side of the border. And the bilingual treasure of the peninsula is another asset. In areas close to the borders people officially use two languages, either Croatian–Italian or Slovenian–Italian and in the former Yugoslav parts of the peninsula, Italian is considered the second language, inferring that children learn it in school. The peninsula's common Italian heritage is thus integrated into the everyday life of most people in the region.

Hence, if it was not for the presence of the borders as well as the turbulent history of the peninsula, it would not be characterized by the many differences that make life attractive here, not just to the local population but also to the many visitors coming to the peninsula every year. Tourism, upon which the income of most people on the peninsula depends, is made possible by the tolerant, almost careless attitude of locals towards others. The amount of tourists taken into consideration, their presence is felt by the locals, but the differences they import into life every summer is accepted as part of this life, in the same way as their own differences are accepted. This is what life on the peninsula is all about, the presence of many customs, traditions, languages, institutions, nations and people in a very confined space.

Multiple Borders, Multiple Politics

What Istria may illustrate better than in many other European border regions, is how geopolitical decisions and major changes in them interferes in a multiplicity of ways with life lived in a border region. Joras' identity as a Slovene national is threatened because of official politics and a reluctance at this level to confront the issue of finally determining where the border is to be drawn. The service workers, but also the fishermen in Umag feel that they are left out of the picture when the political system negotiates matters that concern their very existence. For Karlo-Carletto who is Croat in the morning and Italian at night, the border is an integral part of life and, especially, of his sense of identity. And many people on Istria just want to go on living a life where the border is not made into a hindrance of interaction but remains an asset to it – very unlike those for whom the border is an abstract construction articulating differences between stereotypical images of national identities.

Borders are performed or enacted at various levels, including geopolitics, public and political discourse, historical narrative and everyday life practices. Borders exclude and include, at the same time as they open up towards sensations of a space of total inclusion, absolute hospitality. Borders are functional and symbolic at once, they help us to construct national, regional, local and European identities. Borders generate conflicts between people and sometimes they even generate conflicts within people. Borders help interaction and exchange between people with different backgrounds, in the form of capital or in the form of experiences. But foremost, borders prove that culture and politics always connect in the concepts we use to articulate who we are, concepts we also act upon. In other words, our mundane, every day practices making borders visible, provide a key for us to understand the political enactment and performance of borders. Policy approaches focusing on cross-border partnerships and cooperation in Europe, tend to forget such practices and interactions producing borders and how such practices play a constitutive role in consolidating the very existence and non-existence of borders.

References

Andersen, D.J. (2011), 'Living in the Shadows of Geopolitics – Everyday Practices and Public Discourse at the Croatian–Slovenian Border', *Centre for International Border Research (CIBR) Electronic Workingpaper Series*, www.qub.ac.uk/cibr

Barbic, A. (2004), 'Perceptions of New Realities along the Slovenian–Croatian Border', in Pavlakovich-Kochi, Morehouse and Wastl-Walter (eds),

Challenged Borderlands – Transcending Political and Cultural Boundaries, Aldershot: Ashgate.

BBC news (2009), Croatia Slovenia border row heats up: http://news.bbc.co.uk/2/hi/europe/7896040.stm

Berg, E. and Ehin, P. (2006), 'What Kind of Border Regime is in the Making? Towards a Differentiated and Uneven Border Strategy', *Cooperation and Conflict* 41(1): 53–71.

Boedeltje, F.J., van Houtum, H. and Kramsch, O.T. (2006), 'Communication from the European Union: The European Neighbourhood Policy (ENP) "The Latest Addition to our Democratization Toolbox!"' 'EUDIMENSION' STATE OF THE ART STRATEGY PAPER. Nijmegen.

Clifford, J. and Marcus, G. (eds) (1986), *Writing Culture: The Poetics and Politics of Ethnography*, Berkeley: University of California Press.

Donnan, H. and Wilson, T.M. (1999), *Borders: Frontiers of Identity, Nation and State*, Oxford: Berg Publishers.

Drakulic, S. (1996), *Cafe Europa*, London: Penguin Books.

Eder, K. (2006), 'Europe's Borders: The Narrative Construction of the Boundaries of Europe', *European Journal of Social Theory* 9(2): 255–71.

European Commission (2006), *Regional Cooperation in the Western Balkans. A Policy Priority for the European Union*, Luxembourg: Office for Official Publications of the European Communities.

Galasinski, D. and Meinhof, U.H. (2002), 'Looking Across the River: German–Polish border Communities and the Construction of the Other', *Journal of Language and Politics* 1(1): 23–58.

Garfinkel, H. (1967), *Studies in Ethnomethodology*, Englewood Cliffs, NJ: Prentice Hall.

Hinchliffe, S. (1999), 'Technology, Power, Space – The Means and Ends of Geographies of Technology', *Environment and Planning D: Society and Space* 14: 659–82.

Kramsch, O. (2002), 'Navigating the Spaces of Kantian Reason: Notes on Cosmopolitan Governance within the Cross-border Euregions of the European Union.', *Geopolitics* 6(2): 27–50.

Krnel-Umek, D. (2005), 'Slovenians on the Adriatic', in *The Slovenian–Croatian border in Istria – Past and present*, Ljubjana: Narodna in univerzitetna knijznica.

Krnel-Umek, D. (ed.) (2005), *The Slovenian–Croatian Border in Istria – Past and Present*, Ljubjana: Narodna in univerzitetna knijznica.

Lampe, J.R. (2000), *Yugoslavia in History – Twice There was a Country*, Cambridge: Cambridge University Press.

Latour, B. (1993), *Reassembling the Social: An Introduction to Actor-Network Theory*, Oxford: Oxford University Press.

Latour, B. (1986), *We Have Never Been Modern*, Cambridge MA: Harvard University Press.

Laurier, E. (2011), 'The Region as a Socio-technical accomplishment of Mobile Workers.' Online paper, Glasgow University, www.web2.ges.gla.ac.uk/~elaurier/texts/region.pdf

Ljubljana Life.com (2008), 'Trouble at the Border'. http://www.ljubljana-life.com/ljubljana/border-dispute [accessed 30.08.2010].

Löfgren, O. (1999), 'Crossing Borders – the Nationalisation of Anxiety', *Ethnologica Scandinavica* 29: 5–27.

Meinhof, U.H. (ed) (2002), *Living (with) Borders: Identity Discourses on East–West Borders in Europe*, Aldershot: Ashgate.

Meinhof, U.H. and Galasinski, D. (2002), 'Reconfiguring East–West identities: Cross-Generational Discourses in German and Polish Border Communities', *Journal of Ethnic and Migration Studies* 28(1): 63–82.

Mol, A. (2002), *The Body Multiple. Ontology in medical practice*, Durham, NC: Duke University Press.

Paasi, A. (2005), 'The Changing Discourses on Political Boundaries – Mapping the Background, Contexts and Contents', in H. Van Houtum, O. Kransch and W. Zierhofer (eds) *B/ordering Space*, Aldershot: Ashgate, pp. 17–31.

Pavlakovich-Kochi, V. (2004), 'The Zagreb Region as Double Doorsteps: Implications of a Redefined Boundary for Regional Development', in V. Pavlakovich-Kochi, B.J. Morehouse and D. Wastl-Walter (eds), *Challenged Borderlands – Transcending Political and Cultural Boundaries*, Aldershot: Ashgate, pp. 91–108.

Pavlakovich-Kochi, V. and Stiperski, Z. (2004), 'The Croatian–Slovenian Border: The Local Experience', in V. Pavlakovich-Kochi, B.J. Morehouse and D. Wastl-Walter (eds.): *Challenged Borderlands – Transcending Political and Cultural Boundaries*. Aldershot: Ashgate, pp. 237–50.

Perkmann M. (2007), 'Policy entrepreneurship and multilevel governance: a comparative study of European cross-border regions', *Environment and Planning* C 25: 861–79

Rumford, C. (2008), *Citizens and Border Work in Contemporary Europe*, London: Routledge.

Rumford, C. (2006), 'Theorizing Borders', *European Journal of Social Theory* 9(2): 155–69.

Sahlins, P. (1989), *The Making of France and Spain in the Pyrenees*, Cambridge: Cambridge University Press

Sandberg, M. (2009), 'Grænsens nærvær og fravær: Europæiseringsprocesser i en tvillingeby på den polsk-tyske grænse', PhD thesis, University of Copenhagen (in Danish, can be accessed at www.dasts.dk).

Schultz, H., Stoklosa, K. and Aeniak-Quast, D.J. (2002), *Twin Towns on the Border as Laboratories of European Integration*, Frankfurt an der Oder: Frankfurter Institut für Transformationsstudien.

Scott, J.W. (2010), 'Building Cross-border Regions: Lessons from Europe on the Role of Civil Society', Conference paper: 2010 PCRD International Conference.

Scott, J.W. (ed.) (2006), *EU Enlargement, Region Building and Shifting Borders of Inclusion and Exclusion*, Aldershot: Ashgate.

Walters, W. (2006a), 'Border/Control', *European Journal of Social Theory* 9(2): 187–203.

Walters, W. (2006b), 'Rethinking Borders Beyond the State', *Comparative European Politics* 4(2–3): 141–59.

van Houtum, H., Kramsch, O. and Zierhofer, W. (eds) (2005), *B/ordering Space*, Aldershot: Ashgate.

Zaiotti, R. (2011), 'Performing Schengen: Myths, Rituals and the Making of European Territoriality Beyond Europe', *Review of International Studies* 37: 537–56.

PART III
Cross-border Cooperation and Mobility

Chapter 8
Boundary Objects in Border Research: Methodological Reflections with Examples from Two European Borderlands

Jouni Häkli

Introduction

International boundaries are deceptively simple. As crisp lines on maps or concretely marked in geographical space they communicate clear-cut division between distinct state territories, between here and there, the inside and the outside. Yet this geometric clarity is superimposed upon multiple complexities that characterize international borderlands. While boundaries seldom separate two linguistic or cultural realms in any precise manner, it is commonplace to find international borderlands as amalgamations of two or more intermingling sets of linguistic and cultural resources, such as traditions, habits, historical narratives and customary ways of doing and understanding things. Moreover, decades and even centuries of state-based socialization through school education, political participation and the media, have resulted in borderlands evolving as the meeting points of two (or more) national societies with dissimilar institutional, legal and political systems, partly separate local economies and at times even different currencies.

On top of this socio-cultural and institutional messiness, there are different ways in which boundaries come to be enacted by people involved in dissimilar social positions, roles and settings. Hence, as is the case in Catalan borderlands between Spain and France, for those involved in Catalan nationalist activities and organizations the boundary represents a violation of the territorial integrity of 'Greater Catalonia' (Els Països Catalans), whereas for the practitioners of official cross-border cooperation the boundary is a meeting point and motive for cooperation and, as such, a legitimate element in the border landscape (Häkli, 2001).

Boundaries are an inextricable part of cross-border cooperation. However, it is fair to state that international borderlands are extremely challenging contexts for pursuing coordinated action. Several studies have shown that cross-border cooperation is a continuous struggle to bridge actors across dissimilar socio-cultural and institutional contexts, various degrees of authority and political autonomy and distinct systems or legal regulation. Hence, at stake in cross-border cooperation is a complex governance effort tackling the challenge of coordinating expertise by 'communities of practice' embedded in different socio-cultural contexts (e.g. Gualini, 2003; During et al., 2007; Perkmann, 2007; Leibenath et al., 2010).

To make sense of this messiness, and some means to reduce it, this chapter explores the potential of the concept of boundary objects in the analysis of cross-border cooperation. Empirically the chapter deals with the complexities related to the different enactments of the international boundary between Spain and France in Catalonia. First the idea of boundary objects is outlined with focus on factors that maximize both communication and autonomy in interactions between communities of practice. The chapter will then move on to arguing that international boundaries perform the functions of boundary objects and that this accounts for their pervasiveness even in contexts where the freedom of mobility across boundaries has been a defining feature of the development of the borderland over the past decades, such as the Catalan border landscape. The chapter concludes by showing that boundary objects may be consequential for alleviating the socio-cultural complexities related to borderlands and especially to successful cross-border cooperation in Catalonia and beyond.

The Idea of Boundary Objects

The notion of boundary objects was coined by Susan L. Star and James R. Griesemer (1989) in the context of an ethnographic and historical study of scientific work conducted in the Museum of Vertebrate Zoology at the University of California, Berkeley in the early 20th century. That a natural history museum would come to host scientific research is not self-evident because these institutions were often established as exhibitions to evoke popular wonder and to provide public instruction, rather than to serve as contexts for scholarly contemplation on research problems. What Star and Griesemer (1989), therefore, found interesting is precisely why and by what means did the Museum of Vertebrate Zoology from its very inception develop as a museum devoted to scientific research. To this end they studied historical records concerning the administrators of the museum, the amateur naturalists who contributed specimens to the museum, the professional biologists conducting research in the museum, and other actors forming the network that made the scientific work possible (philanthropists, conservationists, preparators and taxidermists, and the general public). Their goal was to solve the puzzle of how the network succeeded in crafting a coherent problem-solving enterprise despite the fact that it brought together very different social worlds of amateurs, bureaucrats, professionals and scientists, and required cooperation beset by a challenging heterogeneity.

In accounting for this successful cooperation Star and Griesemer explored the different visions that stemmed from the participating social worlds and in particular paid attention to how the different 'communities of practice' could communicate and work together without consensus (Star, 2010). It is here that Star and Griesemer (1989) propose the concept of boundary objects as referring to any concrete or abstract element that people can use as a point of reference in their interactions. Boundary objects work to maximize communication without requiring consensus between the interacting parties. This sets two simultaneous demands for them. First, they have to be characterized by 'interpretive flexibility' so as to accommodate communication across differences. Second, in order to retain significance and attractiveness they have to be specifiable through refinements in the context of particular locations and practices so as to secure the autonomy of the cooperating parties. Boundary objects are, thus, 'both ambiguous and clear, at different moments, for different purposes' (Star, 2002, p. 118).

> Boundary objects are … both plastic enough to adapt to local needs and the constraints of the several parties employing them, yet robust enough to maintain a common identity across sites. They are weakly structured in common use, and become strongly structured in individual-site use. These objects may be abstract or concrete. They have different meanings in different social worlds but their structure is common enough to more than one world to make them recognizable, a means of translation. The creation and management of boundary objects is a key process in developing and maintaining coherence across intersecting social worlds. (Star and Griesemer, 1989, p. 393)

In highlighting interpretive flexibility the notion points at a least common denominator recognized by the members of different social worlds. This, however, does not lead to a consensus-driven process of communication and action but rather the opposite: it provides an understanding of how cooperation may benefit from objects that can be shaped by all parties according to their needs. Hence, as 'ill structured' the object (at once material and processual) resides between social worlds, but different groups can at once keep working on the object making it more specific and tailored for their own specific purposes and thus 'useful for work that is not interdisciplinary' (Star, 2010, p. 605). It is this tacking back and forth between flexible and rigid interpretation of an object that enables cooperation without consensus as a particular kind of problem solving enterprise.

Importantly for the purposes of this chapter, Star (2010, p. 605) stresses that when the oscillation between the two forms becomes standardized, then boundary objects begin to change into infrastructure, into standards, things and yet other processes. This is how cooperation across international boundaries, broadly understood, may gradually become institutionalized, turning the boundary from versatile ill-structured objects to standardized infrastructure and resource for further transnational interactions. Thus, as a consequence of intensifying cooperation the boundary will not weaken or disappear but rather it turns into a durable resource for a growing number of cooperative efforts. I elaborate this development further below.

Through an analysis of specimens, methods, field notes and maps related to the emerging research activities in Berkeley Museum of Vertebrate Zoology, Star and Griesemer (1989) identify four types of boundary objects that are both concrete (material) and abstract (codified, symbolic) in their use. 'The repository' is a set of indexed objects grouped in a standardized fashion, such as a museum or library. What makes repository a boundary object is the way it enables communication and action across heterogeneity by means of modularity: 'People from different worlds can use or borrow from the 'pile' for their own purposes without having directly to negotiate differences in purpose' (Star and Griesemer, 1989, p. 410). If we think of repository in the context of cross-border cooperation, common databases established for representing and analysing cross-border phenomena – notoriously difficult to depict due to differences in nationally-based statistical systems – stands out as an apt an example. A case in point is the Interreg IIIA project ECCOMAP that aims at establishing cross-border databases between Extremadura Autonomous Region in Spain and the Alentejo Region in Portugal (ECCOMAP, 2011).

'Ideal type' is another kind of boundary object that functions to overcome local or singular specificities through abstraction. It thus serves as an effective means of symbolic communication and cooperation because as abstractions they remain fairly vague and thus adaptable to local needs. As examples Star and Griesemer (1989, p. 410) give diagrams, atlases or other descriptions that do not accurately detail the location or thing they depict. In the context of cross-border cooperation ideal type could refer to any symbolic use of maps to highlight, for example, the unity of the cross-border region (e.g. Häkli, 2009).

'Coincident boundaries' refer to objects that are identical in regard their contours but have different internal contents. As an example of coincident boundaries Star and Griesemer (1989, p. 411) give the territorial shape of the state of California, which enabled the amateur naturalists, the administrators, the scientists to conduct their work autonomously with a firm understanding that all this work, carried out by different participating communities of practice, actually deals with the same object, California. In cross-border cooperation coincident boundaries are recurrently resorted to in multiple ways, both abstractly and concretely. For example, when land use in a cross-border region is planned as a joint effort, it may be necessary to carry out two separate zoning processes due to different jurisdictions that regulate the processes in different sides of the international boundary. Here cartographic alignment of the zoning areas enables the two processes to be carried out autonomously but synchronizing the outcome by means of a joint boundary linking the two planning areas (Häkli, 2009).

'Standardized forms' are the fourth type of boundary objects Star and Griesemer (1989) identify in their study. The term refers to methods for reducing the uncertainties pertaining to work carried out in dispersed locations and heterogeneous groups. If we extend the idea of standardized forms to include the regulatory setting under which cross-border cooperation is undertaken, the European cooperation instrument European Grouping for Territorial Cooperation (EGTC) provides an illuminating example. EGTC is a treaty based legal instrument established in 2007 at European Community level to overcome the obstacles hindering territorial cooperation. The instrument enables the creation of cooperative groupings formed by states, regional or local authorities, associations or any other public bodies in Community territory. Importantly the EGTC grouping has legal personality, which simplifies the management procedures of cross-border cooperation as these no longer need to be established separately for all cooperating legal entities but the EGTC can be the sole responsible for project management (INTERACT, 2011). The standardization of cross-border governance by means of EGTCs is of importance to all forms of cooperation that have suffered from complexities related to having to operate under different national legislative frameworks. Interestingly, the Euroregion Pyrenees-Mediterranean was officially established as one of the first EGTCs in August 2009 (Euroregion, 2011).

In presenting the idea of boundary objects at work in the Berkeley Museum of Vertebrate Zoology, Star and Griesemer (1989, p. 410) are clear to point out that the four types and the empirical examples they give are not an exhaustive list in this case, let alone more broadly. They propose the types as reflective of analytical distinctions within a system of boundary objects, not as ideal types that can be found in all contexts. Indeed, some characteristics of the types of boundary objects seem to overlap and a single object could be seen to embody two or more of these types (as in the case of ideal type and coincident boundaries, which both could apply to territorial representations of space as boundary objects).

Given the 'interpretive flexibility' of the concept itself, it is not surprising that there are numerous different uses that the concept has been put in scholarship following Star and Griesemer's seminal work. The notion has found foothold especially in areas of research focusing on expert systems and collaborative scientific work (e.g. Aibar and Bijker, 1997; Henderson, 1998; MacEachern, 2001; Jensen, 2005; Prasad, 2007; Schneider, 2009). While criticisms have been levelled against simplistic or anecdotal uses of boundary objects as referring to any artefact involved in coordinated actions, there is a sizeable literature portraying nuanced analyses of collaborative processes and taking seriously the complexity of the interactions between social worlds (Trompette and Vinck, 2009).

Somewhat concerned about the at times overly flexible uses of the notion, also Susan Leigh Star (2010) has in retrospect proposed some limits to the analytical application of boundary objects concept. While refraining from normative statements about the true meaning and use of the concept, she nevertheless suggests that boundary objects are analytically most useful when applied in proper scale and scope. In regard the former, the notion is best at home when it is used to analyse processes occurring at organizational level. Hence, whatever the context it is employed in, it is the organization of collaborative action and the intersection of participating worlds at stake in the process that can be better understood through the lens of boundary objects (Star, 2010, p. 612).

When it comes to the proper scope of the notion, Star (2010) suggests that it is best seen as specific to collaborative efforts rather than applicable to any artefact that portrays interpretive flexibility (such as a flag, the Bible or a particular film). However, Star (2010) is careful to stress that, for example, an American flag might well play a role in a collaborative effort to advertise, market, and distribute flags.

But again, the proper scope of the notion restricts its use to organizational level and the role that artefacts play in this domain (Star, 2010, p. 613).

Mindful of these conceptual limits, I propose that the notion of boundary objects may offer interesting perspectives and novel approaches to the study of multifarious cross-border cooperation processes. First and foremost I think that the notion is valuable as a methodological orientation that helps pay attention to practices instead of focusing on structural and institutional conditions alone. Moreover, the approach brings to the fore the role of material objects and artefacts that may be consequential for cross-border cooperation, including the materiality of the international boundary. This may be helpful in bringing into balance the constructionist and materialist readings of boundaries as elements of everyday political and cultural landscapes. Thus, seen in terms of boundary objects, the international boundary is multiple. It is both abstract and concrete, general and specific, conventional and unforeseen, process and a protocol. It has different meanings in different worlds, but those meanings are sufficiently structured to be mutually recognized. As a methodological guideline, boundary objects approach points at the question of how actors involved in cross-border cooperation maintain their differences and their cooperation, how they manage and restrict variety, and how they coordinate in space and time. It is a conceptual tool highly attentive to particular contexts and thus well-tuned for empirical work focusing on different kinds of borderland contexts. It makes visible the ways in which actors operating in socially and spatially contingent circumstances come to establish and maintain coherence between intersecting social worlds, without an overarching consensus as the condition of their collaborative work.

Boundary Objects in Cross-border Cooperation

The notion of boundary objects has been used in a wide variety of research areas including research on collaborative systems, organizational change and social studies of science and technology. Perhaps surprisingly the notion has found only few applications in border studies, despite the fact that the concept of boundary object would appear to resonate fairly well with the conceptual repertoire of this multidisciplinary research area. The few works that use the concept of border objects mainly employ the notion as a heuristic device that might explain the success (or failure) of collaborative work across international boundaries (e.g. Grygar, 2009; Leibenath et al., 2010; Wilder et al., 2010). While in these works the idea of boundary objects is used in its proper scale and scope, they are not in-depth analyses designed from the outset to apply the methodology of boundary objects approach.

In view of this apparent research lacuna I wish to probe further with the idea of boundary objects as one potent methodological guideline in the analysis of cross-border cooperation. I do not propose that this approach is once-and-for-all solution to all analytical challenges involved in understanding and explaining cross-border phenomena, but rather wish to outline relatively strict limits to where the best potential of boundary objects might lie in this research field. To argue for the added value of this approach I propose some research questions, or 'puzzles', where its potential may have analytical purchase in the scrutiny of cross-border cooperation. While the discussion will use examples from the cross-border cooperation at the Catalan borderlands between Spain and France, I will also resort to my previous experience with a research project on Swedish–Finnish cross-border cooperation, where I first sought to develop the idea of boundary objects as a factor in transnational collaboration.

To begin to narrow down the scale and scope of the areas of application for the boundary objects notion, it is useful to take heed of Susan Leigh Star's (2010) suggestion that the concept works best when applied in the kinds of research settings it was initially designed for. To this end, the analytical focus of boundary objects analysis should lie in a relatively coherent collaborative effort where the *key actors can be known* and even named. In the case of cross-border cooperation this is not uncommon and thus the limitation will not restrict the viable use of the approach too much. Importantly, there is more at stake here than just locating the key actors as a matter of fact. Because the boundary objects approach is well suited for ethnographic analysis, the very procedure through which the key actors is traced may be a significant part of the emerging understanding of what the collaborative network is like, where its 'system boundaries' lie, how these boundaries are produced and what the role of amateurs and lay people is in the unfolding cooperation. In other words, the methodological rule of focusing on key actors does not rule out major interest in the participation, or non-participation, of the ordinary borderlanders in the cross-border cooperation. In fact, paying attention to the 'system boundaries' through which the key actors coordinate and control their collaboration may be highly consequential for the understanding of why certain collaboration is successful or fails to be that.

Targeting specifically the organizational level of social action brings us to the second limitation proposed by Star (2010): To remain in its proper scope the boundary objects analysis should focus on a joint effort to accomplish something, rather than analyze any intersection of heterogeneous social worlds that might have a relation to cross-border cooperation. Again, this is typical to cross-border cooperation and thus the condition will not preclude the approach from being a useful analytical tool in the research field.

Together the two methodological rules appear to lay the approach at the very centre of research on cross-border cooperation. After all, these processes tend to be purposeful efforts to accomplish clearly articulated goals through a collaboration of certain players representing organizations, institutions and communities from both sides of international boundaries. For example, in the context of Catalan borderlands there are numerous ongoing collaborative efforts to foster local and regional development (e.g. Four Motors for Europe network), develop public service provision (e.g. the Cerdanya cross-border hospital), institutionalize incipient forms of governance (e.g. Euroregion Pyrenees-Mediterranean), to name just a few (Generalitat, 2011). In any of these cases it would be possible to determine who the most important actors in charge of the cooperation are, what their shared and specific goals are, which areas of professional expertise they represent and how their worlds intersect in the collaborative processes.

Yet, there certainly are questions that the methodology works to solve better than others. To trace these it is good to recall the ethnographic origins of the boundary objects approach as developed by Star and Griesemer (1989). Star (2010) mentions several previous research endeavours that contributed to the idea of boundary objects in the analysis of scientific work in Berkeley Museum of Vertebrate Zoology. What these all have in common is the goal of looking at scientific practice as it unfolds 'on the ground' – in laboratories and through scientists' interactions. At stake here is less a specific, narrowly defined approach to on-site field work or the collection of empirical data by means of participant observation. Rather the defining element in boundary objects methodology is that it encourages the researcher to ask unconventional questions and seek answers from events and phenomena that might be disregarded as irrelevant for the processes under scrutiny. In other words, the approach at its best is a way of 'anthropologizing' cross-border cooperation by directing attention also to actions, elements and objects that the actors themselves cannot account for as important but which they nevertheless use routinely to carry out their work.

While on-site observation of collaborative practices may be a relevant research strategy, it is important to remain open-minded when it comes to choosing of empirical materials for in-depth analysis. Star and Griesemer (1989) themselves mainly resort to historical records and collections of specimens from which they trace aspects that they interpret as performing the function of boundary objects. In a similar vein I analysed the development of cross-border cooperation since the late 1990s in the twin towns of Haparanda and Tornio at the boundary between Sweden and Finland (Häkli, 2009). Methodologically I based my analysis on various documents related to the planning and realization of the *På Gränsen – Rajalla* cross-border cooperation project. The project (literally meaning 'on the border') is a still ongoing effort at coordinated community planning between Haparanda and Tornio with the aim of physically uniting the two towns by constructing a common town centre.

My aim was to look at how different communities of practice have pursued cross-border cooperation bridging the social worlds created through distinct areas of professional expertise and national discourses and practices. From documents and publications related to the *På Gränsen – Rajalla* project I traced the ways in which the Tornio River, a 'natural boundary' between Sweden and Finland, was crafted as a boundary object in cooperative networks. While the materials I analysed did carry certain limitations, including their emphasis on the visual and verbal at the cost of direct observation, they nevertheless enabled me to approach, in an unconventional vein, the means that have been utilized to facilitate communication and cooperation across intersecting social worlds at this international borderland.

This effort resulted in a close re-reading of the project documentation especially in view of how the Tornio River was adopted as a three-fold boundary object. First, the Tornio River was significant as a physical entity vested with a degree of agency in relation to the project. Second, the representation of the Tornio River was important as a means of overcoming the dividing function of the border especially from the point of view of physical planning. Third, the river acquired a role as a symbolic space in the border landscape, thus helping to extend the sphere of the project from the core actors toward the broader citizenry. The national boundary performed as the river's affective kernel and locus of passion – as spatial symbolism that helped unite differing experiences between professional worlds and the borderlanders' lived situations (Häkli, 2009).

A similar anthropologizing approach might open up new understandings of what makes cross-border cooperation successful in Catalonia. Some factors in this regard are already well established, such as the fact that both Spain and France have a Catalan speaking minority living at the Catalan borderland, even though the political and economic position of the Spanish Catalonia, *el Principat*, clearly exceeds that of its northern counterpart (Mansvelt-Beck, 1993; Mancebo, 1999). There also exists a strong cultural affinity across the French–Spanish boundary supporting cross-border cooperation. Adding to this, both Spain and France are Schengen countries, which makes the cross-border mobility of goods and people smooth and easy. Given the fact that Catalonia is one of the more active European regions with a marked interest in forging relations of cooperation with partners in other European countries, it is hardly surprising that examples of successful projects are many, ranging from sister cities' exchange of experience and information to local cross-border cooperation in areas such as the annual vehicle inspection, water treatment and the development of tourism (Guibernau, 1997; Roig, 1997). Larger scale projects range from the mountain regions' cooperation under the Working Community of the Pyrenees (*La Comunitat de Treball dels Pirineus*) to a programme for establishing the Catalan Cross-Border Eurodistrict (*Eurodistricte de l'Espai Català Transfronterer*) as a legal entity to support cooperation aimed social, cultural and political development of the Catalan borderland (Generalitat, 2011).

Yet, as is the case with most collaborative work involving experts with different backgrounds and areas of specialization, Catalan cross-border cooperation is not without challenges. Adding to this heterogeneity is the nature of cross-border cooperation as professional activity not confined in laboratories, museums or the pages of professional journals. On the contrary, the collaboration and its results are to a great extent a public matter involving the views, opinions and sometimes even participation by the broader citizenry. As has been shown in several studies of cross-border cooperation, these broader ramifications are far from simple or predictable when it comes to acquiring acceptance and legitimacy for this collaborative work and its goals (Häkli, 2001, 2002; Strüver, 2004). Hence, in Catalonia the contemporary social and economic conditions under which various forms of cross-border cooperation take place include different perspectives on the border and its meanings. Those who are institutionally involved in official cross-border cooperation tend to view the borderland differently from those who are committed to, say, nationalist goals. The rather extreme views of the latter, again, may differ considerably from those of the borderland denizens, for whom the borderland merely represents an everyday environment marked by certain physical and social characteristics (Häkli, 2001).

It is for assessing the terms under which cross-border cooperation succeeds both as a professional effort and in gaining political legitimacy through popular acceptance that the boundary objects methodology might prove useful. As I have argued above, successful cross-border cooperation is similar to scientific work in that it may utilize boundary objects to help in negotiating and resolving potentially conflicting sets of concerns that arise from the intersection of the multiple communities of practice involved. To this end they function as intermediaries between social worlds that should be able to communicate, cooperate and stay committed within this heterogeneous social landscape.

Enacting the French–Spanish Boundary as Boundary Objects

As shown in the discussion above, boundary objects can be of many different kinds. However, in the case of Catalan borderlands there is one particular aspect to the French–Spanish border that stands out as providing an interesting example of the potential of this methodology. As is well known, the border is demarcated along the highest elevation of the Pyrenees, a monumental physical divider between Spain and France. Yet, instead of merely separating people the mountains also represent a cultural link between the Spanish and French Catalonias, a cross-border region inhabited by 'mountain people' (Häkli, 2004). By means of boundary objects approach it is possible to grasp the intertwinement of social and material realities, instead of working merely within the constructionist realm of cultural signification. The latter emphasis points at one of the intellectual homes of the boundary objects approach: the Actor Network Theory (ANT).

Besides deriving from ethnographic research orientation within science and technology studies, the boundary objects methodology was developed in the intellectual context of ANT. From the outset the proponents of ANT have considered human and non-human actors and social and material elements as potentially equally relevant to the understanding and explanation of social phenomena. In setting forth the concept of boundary objects Star and Greisemer (1989) wanted to develop further the ANT model of 'translations' as developed by Bruno Latour, Michel Callon and John Law. They saw that ANT has overemphasized the perspective of an aspiring scientist or other entrepreneur who attempts to enrol allies by re-interpreting their concerns to fit their own programmatic goals. Instead of such mono-perspectival accounts of translation Star and Griesemer proposed that it is necessary to analyse the multiple negotiations that are needed for cooperation occurring at the intersection of different participating social worlds (Star and Griesemer, 1989, p. 396).

To accept that material artefacts and elements are potentially important as boundary objects it is not necessary to digest all claims of ANT, such as the symmetry between human and non-human agency. Indeed, many scholars interested in the approach may find it more feasible to work with a 'weak version' of ANT that views the networks of actors as both social (human) and material (non-human), but not in equal measures because it is social relations that direct the shaping of the networks (Murdoch, 1998; Castree, 2002). This is my own understanding as well. Human actors are 'interactive' while non-human actors are 'indifferent', i.e. unlike the latter, human actors are conscious about their position in the structure and environment around them (Hacking, 1999).

This said, viewing the Pyrenees mountain range as a boundary object may be illuminating of both how the material landscape supports cooperation and how this cooperation, broadly understood, consolidates the international boundary as its standardized 'infrastructure' (cf. Star, 2010). To trace the different guises in which the Pyrenees may function as boundary objects, it is useful to examine the different ways in which the international boundary is enacted by those involved in cross-border cooperation. It is likely that the French–Spanish boundary of the Pyrenees figures simultaneously as a cultural, material and representational reality. As an environment challenged by scarce means of communication as well as other aspects of limited accessibility, the mountains are already part of multilateral negotiation for instance within the perimeter of the Working Community of the Pyrenees that focuses on development issues specific to mountain municipalities. Within this cooperation the mountains are a tangible reality that professionals from different domains can approach as something they consider essentially the same, but nevertheless, are able to specify in ways that reflect their particular framings (e.g. environmental concerns, communications and logistics, boundary effect on regional development, changing demographics, evolving governance).

Cooperation across various areas of expertise is challenging in itself but the role of the Pyrenees as boundary objects extends well beyond professional interactions. Rather than being just a physical matter of fact, the mountains represent various aspects that are essential to Catalan history and identity. On the most fundamental level the Pyrenees are connected to struggles against domination by the Moorish political power in the Early Middle Ages (Nogué, 1998). Later on, resistance toward Castilian centralization policies centred at times on the monasteries of Ripoll and Montserrat, both sheltered by an extraordinary mountain landscape that has become a quintessential symbol of Catalan national identity. Moreover, Catalan poets and writers have depicted the mountains as a sacred and intact virgin nature reflecting the national character of the Catalan people (Nogué, 1998). In the early nineteenth century Catalan intelligentsia were drawn to mountain exploration motivated by nationalist as much as scientific and artistic curiosity. Hiking at the Pyrenees was associated with discovering the Catalan national character and landscape, and by the early twentieth century a hiking association, the *Centre Excursionista de Catalunya*, had become one of the most influential societies of civic and cultural character (García-Ramon and Nogué-Font, 1994).

In all, the symbolic value of the Pyrenees for Catalan identity is considerable. This puts into sharp relief the Treaty of the Pyrenees 1659 that resulted in the annexation of the northern part of Catalonia to France (García-Ramon and Nogué-Font, 1994). Even though it took two centuries before the boundary actually materialized in the mountain landscape, the Pyrenees had irrevocably turned into a landscape that, at once, represented the Catalans' mountain way of life and constituted a natural divider between the southern and northern parts of the Catalan lands.

It is this contested but well established history that may provide for the interpretive flexibility engaging the broader citizenry in the border landscape in which cross-border cooperation is embedded in Catalonia. It is clear that for the professional networks of cooperation the Pyrenees often figures as a cultural link or a bridge composed of various means and forms of interaction. However, in the light of existing scholarship it is less evident that the professionally-driven cross-border cooperation can evolve into a platform for broadly shared political and cultural identity across the French–Spanish border. This, in fact, is not yet commonly found at European borderlands. Instead of a passionate will to exploit the possibilities of the increasingly open EU internal boundaries, many ordinary borderlanders remain indifferent toward the market opportunities 'on the other side' (van der Velde and van Houtum, 2004). Moreover, there typically exists a gap between institutional cross-border activities and the borderlanders' interests toward and knowledge about these activities (e.g. Häkli, 2001; Kramsch, 2002; Sidaway, 2004; Strüver, 2004). The discrepancy between official and popular views may have the consequence of diminishing the value of cross-border cooperation. Thus, whatever the achievements of cooperation at the official level are, these may remain without deeper rooting in the borderlanders' social and political fabric.

For the latter to happen in Catalonia the Pyrenees should emerge as a boundary object inhabiting several communities of practice and satisfying the informational requirements of each of them, that is, a reality that, in the face of missing consensus among the broader citizenry nevertheless provides a common enough rationale to more than one social world to make the collaborative work recognizable. In such case the Pyrenean cross-border cooperation would gain acceptance and legitimacy among the borderlanders, either consciously or subliminally, bridging the gap between professional and lay concerns.

But even then, the status quo is temporary and remains so until the boundary objects are gradually standardized and turn into boundary infrastructures (Star, 2010). The latter term refer to the realities that are naturalized to the point where their stability holds for more than one world simultaneously and, thus, is not commonly called into question. As Bowker and Star (1999, p. 299) aptly point out, 'naturalization means stripping away the contingencies of an object's creation and its situated nature. A naturalized object has lost its anthropological strangeness'. They stress that when an object becomes naturalized in more than one community of practice, it turns into an infrastructure that acts as a resource for social action by allowing for local variation but with sufficiently consistent structure to serve standard procedures to be carried out with systematic tools (forms, statistics, and so forth) (Bowker and Star, 1999, pp. 313–14).

If the Pyrenees function as a boundary object bridging the worlds of various professionals and experts and extending all the way to the multiple everyday worlds of Catalan borderlanders, the international boundary in all its institutional stability, judicial clarity, and material objectivity is best understood as a naturalized object in the sense proposed by Bowker and Star (1999). Cross-border cooperation is by necessity characterized by openness to dissenting views and dissimilar ways of framing the manifold goals, forms and objectives of this collaborative work. Boundary objects may be helpful in alleviating this lack of consensus. However, the French–Spanish boundary itself is an institution beyond questioning. The boundary is permanent even though not eternal or immutable. Even those Catalan nationalists who would wish to eradicate the boundary come to attest to its existence in their intentions and actions. In this regard the institution of the French–Spanish boundary is a 'boundary infrastructure' serving the rise and fall of more temporary boundary objects that make cross-border cooperation possible. Hence, contrary to some expectations the boundary is not challenged by cross-border cooperation but rather it is deeply embedded in this collaborative work. This at least partly accounts for the resilience of international boundaries in the age of cross-border regionalization, cooperation and mobility.

Conclusion

In this chapter I have proposed that the notion of boundary objects and the concomitant methodology can be helpful in assessing in some novel ways the possible successes and failures of cross-border cooperation. By necessity the discussion here is tentative at best, and much more work is needed to develop the methodology in the context of border studies. However, I hope to have accomplished some conclusions regarding this emergent approach.

First, in view of its proper scale and scope, the notion of boundary objects appears well suited for the analysis of cross-border cooperation and related phenomena. Collaboration across international boundaries is a matter that can feasibly be studied on an organizational level, and as an intentional activity that involves heterogeneous areas of expertise and dissimilar groups of actors inhabiting different social worlds. To this much studied field boundary objects methodology helps import 'anthropologizing' approaches that pay attention to the mundane practices of professional actors and lay people. Moreover as a descendant of actor network theory, the boundary objects methodology is sensitive to the role of material environments and artefacts in how social processes unfold. Both research approaches are relatively scarce in the border studies literature.

Second, the notion of boundary objects, coupled with that of infrastructures, may be helpful in making sense of why the vast growth of cross-border activities has so little undermined the institution of international boundaries. By introducing a conceptual vocabulary attuned to emergence and stability existing in tandem, the boundary objects methodology foregrounds aspects in the political, cultural and economic development that have hitherto been obscured by the very conventional language used in describing these major developments in Europe and elsewhere.

Third, while there certainly are several threads in the boundary objects methodology that apply generally, it is also clear that all borderland contexts are to some extent unique and thus demanding specific adjustments to the approach to make it feasible and adequate for the assessment of the phenomenon at hand. In this regard the boundary objects methodology is a useful guideline for designing novel approaches to cross-border phenomena, rather than a toolbox readily applicable to making sense of cross-border cooperation.

In all, the notion of boundary objects deserves more careful examination and also critical interrogation before its full potential in border studies can be assessed. On the basis on this tentative discussion, the potential certainly is there. The challenges that this theoretically inspiring and empirically productive work presents, I trust, will intrigue scholars interested in culturally and politically oriented border studies.

References

Aibar, E. and Bijker, W. (1997), 'Constructing a City: The Cerda Plan for the Extension of Barcelona', *Science Technology Human Values* 22(1): 3–30.

Bowker, G. and Star, S. (1999), *Sorting Things Out: Classification and Its Consequences*. Cambridge, MA: The MIT Press.

Castree, N. (2002), 'False Antitheses? Marxism, Nature and Actor Networks', *Antipode* 34(1): 111–146.

During, R., Pieters, E., Bergoglio, S., Bonagurio, D., Vettoretto, L., Gelli, F., Caputo, A., Ghiraldini, D., Vani, M., Knieling, J., Othengrafen, F., Papioannou, A., Gorzelak, G., Herbst, M., Krok, K., Olejniczak, K, Maraidoni, C., Voulgari, M., Karamoschos, A., Argirou, C. (2007), 'Cultural Differences in European Cooperation: Learning from INTERREG Practice', Final Report of the INTERACT project CULTPLAN. Available at: http://www.cultplan.org/downloads/RAPPORT%20DEF%202%20LR.pdf [accessed 20.5.2011].

ECCOMAP (2011), Áreas *Prioritarias Para la Gestión de la Biodiversidad,* Available at: http://www1.unex.es/eweb/gic_eccomap/en/ [accessed 5.5.2011].

Euroregion (2011), *Euroregió Pirineus Mediterrània,* Available at: http://www.euroregio.eu/eu/AppJava/cat/ [accessed 10.6.2011].

García-Ramon, M.D. and Nogué-Font, J. (1994), 'Nationalism and Geography in Catalonia', in D. Hooson (ed.), *Geography and National Identity*, Oxford: Blackwell, pp. 197–211.

Generalitat (2011), 'Foreign policy', Available at: http://www.gencat.cat/ [accessed 6.6.2011].

Grygar, J. (2009), 'Devushka and Cigarette: Fluid Migrants across the EU Border', *Slovenský Národopis* 57(5): 546–59.

Gualini,E. (2003), 'Cross-border Governance: Inventing Regions in a Transnational Multi-level Polity', *DISP – The Planning Review* 152(1): 43–52.

Guibernau, M. (1997), 'Images of Catalonia', *Nations and Nationalism* 3(1): 89–111.

Hacking, I. (1999), *The Social Construction of What?* Cambridge, MA: Harvard University Press.

Häkli, J. (2001) 'The Politics of Belonging: Complexities of Identity in the Catalan Borderlands', *Geografiska Annaler B: Human Geography* 83(3): 5–13.

Häkli, J. (2002), 'Transboundary Networking in Catalonia', in D. Kaplan and J. Häkli (eds), *Boundaries and Place: European Borderlands in Geographical Context*, Lanham, MD: Rowman and Littlefield, pp. 70–92.

Häkli, J. (2004), 'Governing the Mountains: Cross-border Regionalization in Catalonia', in O. Kramsch and B. Hooper (eds), *Cross-Border Governance in the European Union*, London: Routledge, pp. 56–69.

Häkli, J. (2009), 'Boundaries of Trust: Building a Transnational Space in Haparanda-Tornio', in J. Häkli and C. Minca (eds), *Social Capital and Urban Networks of Trust*, Aldershot: Ashgate, pp. 205–32.

Henderson, K. (1998), 'The Role of Material Objects in the Design Process: A Comparison of Two Design Cultures and How They Contend with Automation', *Science Technology Human Values* 23(2): 139–174.

INTERACT (2011), *European Grouping of Territorial Cooperation*, available at http://www.interact-eu.net/egtc/egtc/30/16 [accessed 25.5.2011].

Jensen, C. (2005), 'An Experiment in Performative History: Electronic Patient Records as a Future-Generating Device', *Social Studies of Science* 35(2): 241–67.

Leibenath, M., Bluma, A, and Stutzriemer, S. (2010), 'Transboundary Cooperation in Establishing Ecological Networks: The Case of Germany's external borders', *Landscape and Urban Planning* 94(2): 84–93.

MacEachren, A. (2001), 'Cartography and GIS: Extending Collaborative Tools to Support Virtual Teams', *Progress in Human Geography* 25(3): 431–444.

Mancebo, F. (1999), *La Cerdagne et ses Frontières: Conflits et Identités Transfrontalières*, Perpinyà: Editorial Trabucaire.

Mansvelt-Beck, J. (1993), 'Catalan Linguistic Revival at the Franco-Spanish Border', in J. Leerssen and M. van Montfrans (eds), *Borders and Territories*, Amsterdam: Rodopi, pp. 177–92.

Murdoch, J. (1998) 'Of Actor Network Theory', *Geoforum* 29(4): 357–74.

Nogué, J. (1998), *Nacionalismo y Territorio*, Lleida: Milenio.

Perkmann, M. (2007), 'Construction of New Territorial Scales: A Framework and Case Study of the EUREGIO Cross-border Region', *Regional Studies* 41(2): 253–266.

Prasad, A. (2007), 'The (amorphous) anatomy of an invention: the case of magnetic resonance imaging (MRI)', *Social Studies of Science* 37(4): 533–60.

Roig, S. (1997), 'Els Pirineus: una cremallera que es descorda', *Revista de Girona*, no. 181: 29–35.

Schneider, A.L. (2009) 'Why do Some Boundary Organizations Result in New Ideas and Practices and Others only Meet Resistance? Examples From Juvenile Justice', *The American Review of Public Administration* 39(1): 60–79.

Sidaway, J. (2004), 'The Choreographies of European Integration: Negotiating Trans-frontier Cooperation in Iberia', in O. Kramsch and B. Hopper (eds), *Cross-Border Governance in the European Union*, London: Routledge, pp. 173–90.

Star, S. (2002), 'Infrastructure and Ethnographic Practice: Working on the Fringes', *Scandinavian Journal of Information Systems* 14(2): 107–22

Star, S. (2010), 'This is Not a Boundary Object: Reflections on the Origin of a Concept', *Science, Technology and Human Values* 35(5): 601–17.

Star, S. and Griesemer, J. (1989), 'Institutional Ecology, 'Translations' and Boundary Objects: Amateurs and Professionals in Berkeley's Museum of Vertebrate Zoology, 1907–39', *Social Studies of Science* 19(3): 387–420.

Strüver, A. (2004), '"We are only allowed to Re-act, not to Act": Eurocrats' Strategies And Borderlanders' Tactics in a Dutch–German Cross-Border Region', in O. Kramsch and B. Hopper (eds), *Cross-Border Governance in the European Union*, London: Routledge, pp. 25–40.

Trompette, P. and Vinck, D. (2009), 'Revisiting the Notion of Boundary Object', *Revue d'anthropologie des connaissances* 3(1): 3–25.

van der Velde, M. and van Houtum, H. (2004) 'De-politicizing labour market indifference and immobility in the European Union', in O. Kramsch and B. Hopper (eds), *Cross-Border Governance in the European Union*, London: Routledge, pp. 41–55.

Wilder, M., Scott, C., Pineda Pablos, N., Varady, R., Garfin, G. and McEvoy, J. (2010) 'Adapting Across Boundaries: Climate Change, Social Learning, and Resilience in the U.S.-Mexico Border Region', *Annals of the Association of American Geographers* 100(4): 917–28.

Chapter 9

The (in)visible Wall of Fortress Europe? Elite Migrating Youth Perceiving the Sensitive Polish–Ukrainian Border

Anna Gawlewicz and Carsten Yndigegn

Introduction

This chapter investigates the perceptions of the Polish–Ukrainian border developed by young people who cross the border for educational purposes. While the Polish–German border has been heavily researched (Meinhof and Galasinski, 2000a; Schultz et al., 2002; Stoklosa, 2003; Petrat et al., 2009;; Sandberg, 2009), the Polish–Ukrainian border has not been addressed to the same degree by an internationally oriented research. Although the Polish–German and the Polish–Ukrainian borders are deeply bound to the same historical incidents, both border contexts are different (Bialasiewicz and O'Loughlin, 2002; Scott and Matzeit, 2006). After being socially frozen for decades, the Polish–Ukrainian borderland is only slowly starting to develop, and youth mobility – educational and work migration – that first was directed towards more urban areas or towards the old EU-member countries, also directs towards the region's several big cities, which are offering studying possibilities.

We put emphasis on – what we call – elite youth: university and college students who move or migrate to the neighbouring country to develop academically and personally. There are growing numbers of transnational or migrating students both in Poland and Ukraine. Their everyday narratives and views, and together with regular border experiences, offer new research opportunities within a region which due to the recent enlargements of the EU has gained increased attention.

The existing mobility pattern is not balanced. During the research we have observed that Ukrainian youth choose Poland as a study destination more often and more willingly than Polish students choose Ukrainian institutions. For the majority of Ukrainian students Poland as an educational aim is perceived as a chance to achieve better social status and to live closer to *the West*, which offers better study and work opportunities. Polish students most often admit they treat Ukraine as a *second choice* or a security destination – decided on only when not being admitted to a Polish university. Accordingly, we have also noticed that while Poles in Western Ukraine usually decide among very few particular and specialized institutions (the best example here is the Medical Academy in Lviv experiencing a huge interest of Polish youth) and during the study course stay strongly engaged into the national student enclaves, Ukrainian youth in Poland is much more dispersed and prone to integrate.

The Polish–Ukrainian borderland is historically sensitive due to the conflictual past of the region and geographically strategic since Poland's accession to the European Union in 2004 and the subsequent Schengen zone's enlargement in 2007. The intense changes it went through during the last century such as frequent shifts in demarcation, ideological clashes and violent episodes still tend to resonate in collective memory influencing social relations and neighbour awareness.

The chapter introduces the topic which tends to be neglected in the broad body of border literature. Namely, it focuses on youth's personal narratives of lived border experience. It also shows how border is conceptualized as both geopolitical space and a mentally developed collective construction. The border narratives we have gathered reveal how young people perceive and negotiate the existence of a particular border (that is the Polish–Ukrainian border) with all its distinctive features such as everyday border check-point practices, transnational-scale coexistence, imagining and inventing the Other, as well as the EU-related significance. But they also point out how youth conceptualize national borders, what kind of attitudes towards them young people may have, and how emotional the bordering possibly is.

The chapter will analyse and discuss border perceptions of young people in the light of the implementation of the Schengen border, which has negatively been labelled Fortress Europe. It will consider whether the new iron curtain concept finds resonance in the experiences and life-worlds of young students. The analyses cover popular themes present in border discourses and try to investigate whether the introduction of a new border status (the Schengen EU external border) has been reflected in the recognition of the border among migrating youth.

The Traumatic History of the Polish–Ukrainian Border Region

Poland and Ukraine have a long history of coexistence and neighbourhood. Although the states and their statuses changed during vicissitudes of the European history, mixed communities and mutual influences were incessantly present in the region and a unique melting pot of cultures and nations was until the mid-20th century a distinctive feature of the area. Born of relatively integrated, multicultural and multilingual *Galician* Austro-Hungarian Empire, the area stayed a quasi-autonomous region of relative ethnic, cultural and religious tolerance (Chojnowski and Bruski, 2006).

Poland was restored and recognized as an independent state after 1918, while Ukraine was divided between Poland and the Soviet Union. This way, most of the areas of former Galicia were incorporated into Poland, and Ukraine became one of the Soviet states of the USSR. After the Second World War, the border was moved 200 km west which led to Eastern Galicia and neighbouring Volhynia being incorporated into the USSR (Davies, 2008, p. 967; Yekelchyk, 2007, p. 147). The border movement was followed by a radical displacement of the inhabitants in the border region (Kordan, 1997).

In the following decades, cross-border movement or direct contacts across the border were almost non-existent (Wolczuk, 2002; Haase, 2005; Iglicka and Weinar, 2008; Sagan, 2010). Nationalist sentiments as well as communist-era mass-media-wide policy of defaming Ukrainians in Poland and Poles in Ukraine resonated in collective memory and neighbour perception fuelling mutual antipathy, prejudices and stereotypes (Wolczuk, 2002; Haase, 2005; Kasianov, 2006; Copsey, 2008).

At the end of 1980s, petty trade on both sides of the border started to gain popularity gradually allowing the region to re-build social connections and initiating subsequent economic cooperation. The collapse of Communism in Poland in 1989 and the dissolution of the Soviet Union in 1991 resulted in a slowly and patiently constructed 'border of inclusion' between Poland and the now independent state, Ukraine. Poland was the first country to recognize Ukraine's independence in 1991, and since then, Poland continuously stays Ukraine's advocate to the West (Yekelchyk, 2007, p. 196) and – especially – the European Union (Grabbe, 2000; Ziolkowski, 2008; Klatt, 2011). In 1996 an agreement of visa free travel between the two countries were negotiated. Poland's membership of the EU in 2004 and the Schengen area in 2007 drastically limited most kinds of cross-border cooperation and movement, and visas were re-introduced in 2004, at first free of charge, but by 2007 they had to be paid for (Iglicka and Weinar, 2008).

The Multifaceted Construction of Borders

It is acknowledged that bordering extends societal relation building between individuals and the bordering mechanisms are shown in mental representations of basic human interactions (Paasi, 1998). Indeed, Borders are social constructions in geographical space. That said, they are dividing forces between known and unknown, native and foreign, us and them. Accordingly, border regions constitute laboratories in which more widespread interactions among people and ethnic groups can be studied. In border regions people are divided due to political decisions that may be determined as an outcome of a peaceful agreement or a referendum, but more often are the outcome of conflictual and violent interaction on nation-state level. In this respect European borders have been termed 'scars of history'.

Borders are not only social constructions; they produce meaning and significance as well. Paasi has expressed this as the identity producing functions of borders: 'The meanings of boundaries are thus underlined by the fact that identities are produced through these boundaries. They become part of collective identities, shared memories and the sense of continuity between generations. Identities are often represented in terms of a difference between Us and the Other, rather than being something essentialist or intrinsic to a certain group of people' (Paasi, 1998, p. 81).

Borders also determine perceptions of people. As spatial representations of power relations, they become reflected in the minds of the people who live with and along them. Drawing borders imply that mental divisions are created among the populations on each side of the border (Sahlins, 1989; 1998). The mental representations of borders take on a life of their own and persist even after the constitutive conditions of their formation have disappeared (Berdahl, 1999; Meinhof and Galasinski, 2000a; 2000b; 2000c; 2002; 2005; Meinhof, 2002; 2003a; 2003b; Armbruster et al., 2003; Armbruster and Meinhof, 2005). Personal memories related to borders only reinforce this complexity. Young people are also influenced by the historical memories. Although they have no personal experience, stories are transferred from generation to generation. Being young is no vaccine against prejudices and nationalistic tensions. Young people finding themselves in situations where perceptions of the Other are not based on primary experience can develop new kinds of border identities based on personal interaction and contacts in everyday life (Dekker and Jansen, 1994; 1995; du Bois-Reymond et al., 1998). The perceptions represent 'a sense of a national cultural conflict' based on a mutual feeling that the Other is 'a threat to the respective localities and their people' (Jukarainen, 2003, p. 221).

The societal significance of borders has changed historically. The consequences of globalization are reflected in the change of borders in postmodern times. Postmodern borders take on diverse forms, being both permeable and closed. On one side they keep on framing nation state sovereignty, on the other they allow unhindered flow. Openness and closedness follow a complicated pattern that has been compared with a computer firewall that perform intelligent filtering of immigrants being open for the attractive and being closed for the non-attractive (Walters, 2006a; 2006b).

In Rumford's outlook, postmodern borders have become floating corresponding to the flows of the network (Castells, 2000). Postmodern borders turn into what Rumford has termed borderwork. That is to say, borders are being territorially displaced and border controls are in principle being carried out by anyone anywhere. It is the local inhabitants that call the police when they spot an illegal trespasser; it is the all-encompassing surveillance; and it is the flight companies that perform passport control for the immigration authorities at the site of departure (Rumford, 2008).

The Polish–Ukrainian border region calls special attention towards the EU policies of external borders. The EU neighbourhood policy has been developed to further the cooperation between the EU and its neighbours (Boedeltje et al., 2006; Haukkala, 2008; Orbie, 2008; Gänzle, 2009). In contrast to the rhetoric of a borderless Europe and bridge-building across borders that is present in discourses about European integration, the discourse on external EU-borders is loaded with issues concerning social, political, and economic exclusion, as well as closure, isolation and othering. The Schengen border is perceived as a 'tool letting in those who bring wealth (or valuable personal, educational, business, etc. skills) and at the same time keeping out those who bring burden' (Kramsch et al., 2004 cited by

Berg and Ehin, 2006). Quite interestingly, different metaphores such as 'Fortress Europe' (Kofman and Sales, 1992; Klein, 2003; Snyder, 2005), 'gated community' and 'defended neighbourhood' (van Houtum and Pijpers, 2005, 2007) have been proposed to stress the significance and consequences of the EU 'intelligent filtering'. Researchers specializing in eastern European issues emphasize that the continental East–West disparities are being continously strengthened by the Schengen border. As a result a better–worse relationship between western 'winners' and eastern 'second-hand Europeans' is being socially constructed (Hrytsak, 2005), and a new Iron Curtain or Euro-curtain seems to divide the European continent once again (Bialasiewicz and O'Loughlin, 2002; Mrinska, 2006; Szmagalska-Follis, 2009). The proceedings to be accomplished to enable non-EU-citizens access to the Schengen area are costly and time consuming, and it appears as if a paper wall has been built up from bureucracies. The popular concept of Fortress may sound as an exaggerated media buzz word, but the everyday practices and experiences suggest quite the opposite. This article argues that an invisible wall of Fortress Europe indeed does exist, although it may take various forms – often ephemeral and quite different from the classic one the research has accustomed us with.

Research Design, Field Work, and Thematic Analysis

Our research is strongly embedded in a qualitative approach. Empirically, it is based on fieldwork carried out in 2009 among Polish students in Western Ukraine (Lviv) and Ukrainian students in Poland (mainly Rzeszow, Lublin, Cracow). The interviewed sample includes 21 students: 6 Polish students in Lviv, 2 students coming from Polish minority in Western Ukraine and 13 Ukrainian students in Poland. Besides the minority students, the national division is not clear-cut, while some Ukrainian students have weak Polish roots, but come from Ukrainian families and were brought up usually only with Ukrainian language and Ukrainian traditions. The students either come from borderlands or/and decided to start studies within border and neighbouring – post-Galician – areas. Personal and long-term involvement into the border region was the most important criterion for being selected. The sample consists of 16 females and 5 males. Except two older Ph.D. students, they were 19–24-year-olds. The sampling method was a combination of snowball technique and networking within various organizations or university international student offices.

The field-work included in-depth interviews in the Polish language aimed at triggering thematized story-telling and personal narratives. The interviews' length varied from 45 to 80 minutes depending mainly on the interviewee's capacity to supply with narration. A drawing activity was designed to work as an extra prompt for story-telling. The interviewees were asked to draw how they imagine the Polish–Ukrainian border. Vast majority of interviewees prepared a draft image while describing their border associations and stories. The pictures themselves, though simple and rough, represent and reproduce concepts of the multi-dimensional border.

The collected interviews became a subject of a narrative analysis. The analysis enables to grasp the whole narrative potential lurking in qualitative accounts (Earthy and Cronin, 2008, p. 425) and provides a tool to explore linkages, relationships and socially constructed explanations that naturally occur within such accounts (Kvale and Brinkmann, 2009; Maxwell, 2005; Creswell, 2007). Through narration individuals become mirrors for their environment, culture and the whole social context they are embroiled in.

In line with the findings, the following part of the article was divided into three sections. The first two explore the (in)visible wall of Fortress Europe. The last one focuses on Schengen perception and the EU-external-border border recognition.

Visible Wall of Fortress Europe

The predominant part of the border discourse is related to a direct border checkpoint experience. The students usually associate the border only with the checkpoint and their narration tends to be limited to the very act of crossing as well as the procedures it involves. For regular border-crossers the border thus equals the checkpoint and is not extended to other spheres of transnational coexistence. Additionally, the high frequency of crossing the border makes the students notice changes and otherness only in reference to checkpoints. Checkpoints, thereby, become embodiments of a border, while their landscape (control signs, gates, border officers) reflect both the national authority and the passage to a foreign state. The association is the result of the status of the border and specific border regime which includes traditional control as well as the necessity for Ukrainian citizens to possess a valid visa. Crossing the border involves a series of procedures and forces to obey the rules imposed by the regime. Therefore, the checkpoints become the very essence of the border in the classic sense – a symbolic line dividing separate states and national entities – and clearly dominate the narratives.

For migrating students forced to cross the border relatively often, once in 2–3 weeks, the checkpoint assumes the proportions of a specific micro-cosmos – entirely different from the typical national landscapes they are accustomed to. That is to say, it has its distinctive features (the set: buildings with national symbols, control gates, protective fences and nets), its procedures (passport check, personal control), its rules (e.g. particular regulations about products and goods transfer to the other country) and its characters (border police, duty officers and border-crossers). The regular experience strengthens subjective perception of the place and reception of negative elements of the border reality. Checkpoint recollections embrace similar stories of awaiting, discomfort and uncertainty. Young people make here all possible waiting time comparisons often juggling with numbers. These narratives disclose strong feelings: anxiety, anger or even fury. Since crossing the border involves a personal check, also humiliation and mutual suspiciousness become a recurrent theme.

> T12: 'Poles always check on people very long. Everything – if I don't have a criminal record and so on. They have this machine ... But Ukrainians – they just put a stamp ... When you take a train (to Ukraine) they take passports, keep them for two hours and then – give back. But you never know what they actually do with them. Strange … When I leave a train, I feel like a dirty drunkard. The smell. They wake you in the middle of the night at the border … My associations? Showing what's inside your luggage. Throwing your things around by officers which leave you later to clean the mess … A dog which is to check whether you're not smuggling drugs.'

The border (checkpoint) in the youth discourse seems 'a different world', a space where specific rules are in force and every person is obliged to conform – it is an abstract space of bureaucracy, intense investigation, and interrogation. In this sense, border can be understood as an administered organization with clear authority, an institution, transitional and semi-locked space, entering which forces everybody to comply with the series of proceedings and to adjust personal behaviour to the organizational highly bureaucratic culture. This description reminds us to a vast extent of the concept of a *total institution* – structured space cut off from the regular society and social norms where individuals lead enclosed, formally administered rounds of life (Goffman, 1961). Although border as an institution has its obvious differences from the classical idea of total institution (in Goffman's understanding total institution embrace: prisons, mental institutions and psychiatric hospitals, boarding schools, orphanages, nursing homes, monasteries and orders, etc.) there are many overlapping features. We argue that, similar social mechanisms and contracts as well as response practices are developed. They are explored in the following passages.

> I (interviewer): 'What kind of associations you have with the Polish–Ukrainian border?'
> T9: 'My biggest dream is that a bomb is dropped there. A big bomb, really. It's humiliating. Really, humiliating. It's ... Have you ever crossed it?'
> I: 'Yes.'
> T9: 'On foot?'
> I: 'Yes.'
> T9: 'Have you seen those barriers?'
> I: 'Mhm.'
> T9: 'They make me think of those pens for cattle. Or sheep. And they ask you: "Where are you going?!" "What for?!" "Where do you live?!" "Do you have money for a living?!"... [He answers to his previous questions] "It's none of your business!! I live where I want and I study where I want." This thinking is so communist. Rotten communism. Very painful.'

The Ukrainian interviewee quoted above recollects the feeling of anger and humiliation when taking a foot border crossing. In his story the border experience is much exaggerated and very emotional – the border question boosted his irritation. The young man has a drastic 'bomb-dropping' idea, he believes the border should not exist in the present form not only due to the obvious administrative difficulties it causes to individuals, but also because it deprives human beings of their personal rights, humanity, autonomy, self-determination and freedom of movement. That said, he interprets border procedures as denying humanity, reducing human to a subject which becomes only a part of the bureaucratic and structured system. Sudden turns in his narrative (calm and continuous so far), direct, dynamic questions to the interviewer, strong reaction and a role play character of the recollection (in the interview he started to modulate his voice when speaking for a border officer asking formal questions) show the scale of his checkpoint trauma. The interviewee rationalizes the trauma as a communist relict.

Curiously enough, in the student's recollections also other aspects of institutionalized border regime are mentioned: strong division of individuals who participate in border procedures into two strata (border staff and border crossers), clear demarcation of roles (border/duty officers – placed very high on a symbolically imagined hierarchy – versus border crossers – placed low in the imagined hierarchy), physical disconnection of the checkpoint from the world outside (fences, gates, barriers), structured conversation conducted in a very special manner (tone of voice, formal questions), probably increased display of authority symbols (checkpoint landscape and officers' uniforms).

Another prominent example of borderwork in practice comes from a female Polish student in Lviv. She stresses that the specific informal checkpoint culture was developed in a reaction to the border hardening and gradually imposed restrictions. It includes not only smuggling and the whole range of semi-legal practices, but also specific social behaviour: seemingly chaotic, thoughtless and desperate – unacceptable in regular social interactions. Although, the reader must take allowances for the interviewee's exaggerating style, the story may still sound shocking.

> T7: 'There are those petty smugglers at the border. Now I ride with my boyfriend by car ... But I used to cross the border on foot for quite a long time. So ... So, those cigarettes smugglers, those people – they're the dregs of society. And, a-a-and ... you can hear the worst things from them. Complete lack of cultured behaviour. They elbow their way through the crowd (so much). They hit you with their elbows, push you. Once, I remember, when I was coming home ... They'd almost seriously hurt my friend. Border officers just opened the door and they (people at the border) started to push one another making their way through the crowd. My friend was standing there, waiting very close to the (metal) gate. If it wasn't for me – I pushed her and helped her – they'd crash her ribs. She started screaming; I saw what was going on and helped her. Then, she – poor thing – she couldn't catch a breath. That was horrible. Really. You know, there

> are cases of broken bones (at the border). Really – it's a farce! Tragedy. So... *[pause]* this border crossing – it's horrible.'

Considering the border through the prism of the concept of a total institution provides a socio-psychological justification of the behaviour the young woman describes. Individuals exposed to the highly institutionalized 'inhumane' (as they believe) treatment, start perceiving the checkpoint as an absurd bureaucratic space where typical social interactions do not take place and only a deviant behaviour facilitates achieving goals effectively and fast. Literally speaking, being treated inhumane (or – more often – only believing such a treatment takes place), individuals and their social behaviour become inhumane.

According to another interviewee, the people who tend to behave in such a way are 'very specific' and have nothing to do with regular representatives of both nations. They are the *border people* – a group that developed particular behaviour and attitude in a response to the border regime. They consist also of those who consider the border's nearest neighbourhood an opportunity to set a network of informal trade avoiding taxation or simply to earn extra money within the grey market.

> I: 'What do you mean 'specific'?'
> T15: 'It means the people that come there every day, definitely much more often than once in a while – when you cross a border in travelling purposes or so – they are not the same people ... Well, let me say it differently [pause] ... When one comes to the border, he or she experiencing THIS space and THOSE people can develop a false opinion about the whole nation (Ukrainians), well – both nations to be precise. The vast majority of those people at the border consist of petty smugglers (literally: ants), alcohol addicts, all the suspicious "elements" of society, etc. And when someone crosses this border for the first time – this person, hearing the vulgar words all the time, seeing those badly behaving people, can really feel discouraged towards Poles or Ukrainians in general.'

Petty border smugglers open a separate chapter in checkpoint stories. They are an inseparable part of almost each traditional border landscape. That is to say, smuggling becomes a livelihood strategy for many people from the nearest border area. Poles find Ukrainian prices almost Eldorado prices and eagerly smuggle themselves or buy smuggled cigarettes or alcohols. For Ukrainians from marginalized parts of the country, informal trade profit becomes a significant part of a monthly income. Those practices and their scale, apart from the natural market mechanism, reveal also a sad truth about the condition of Ukrainian economy: very low salaries and wages, relatively high unemployment and an unacceptable standard of living within the peripheries of Western Ukraine. A Ukrainian migrating student quoted below is ashamed of the smuggling procedure and feels discomfort facing her compatriots practicing it. But at the same time she tries to justify these efforts.

> T14: 'Everybody knows how this border looks like ... That those ladies ride from Lviv with cigarettes and alcohol. That's horrible. If you take such a bus, you can stay 4–6 hours at the border before everything is checked ... If you travel with Poles and those ladies are also there – you feel like hiding somewhere and not admitting to being Ukrainian. Because, it's horrible – what they do to smuggle cigarettes. But, on the other side – it's their bread, that's how they make a living.'

In a similar vein, a Polish student finds smuggling an obvious part of a border life. In her recollections, border is symbolically represented by the ugly, angry, screaming crowd and smugglers who demand being let into Poland. The interviewee stresses female smugglers, since petty smuggling became especially popular among middle-aged unemployed Ukrainian women:

> T2: '[The young woman laughs] I think it's a huge smuggling border. When I think of this border I see women which have cigarettes' boxes hidden somewhere in their pants … They undress just before the border and put cigarettes wherever it's possible. They use a tape to stick them to their legs. And that's how they want to cross the border. Of course, everybody can see that *she* has something hidden in her trousers. … But she is not embarrassed. At the border they pretend not to see that. They let her go. And there she goes, all like a square. ... A square woman [laughs].'

As it has been shown, among the Polish students the border experience might often become a border trauma since other borders they are used to cross (most often present internal Schengen borders) usually do not enforce participating in institutionalized and long lasting procedures. Accordingly, not being very familiar with aircraft transport they lack similar personal check experiences. Also, Polish youth has been for a vast part of their conscious life bombed with slogans of borderless Europe and free movement. It is not surprising then, they face a major contradiction when leaving Schengen and the EU.

Invisible Wall of Fortress Europe

Due to the intense administrative procedures, the Polish–Ukrainian border plays a role of a distancing and isolating factor in the youth stories. The Polish students could easily (and, apparently, would like to) treat their Ukrainian destination as an extension of their 'own playground'. Unfortunately, the control in checkpoints and long lines to the other side – quite literally – prevent them from feeling the comfort of closeness and regional unity. A Polish student whose home town is situated only some 30km from the border describes her feelings:

> T5: 'This is the worst thing … The worst memories are from the border … We always say: 'Damn, it's so close, a stone's throw. And everything would be so

> great if it wasn't for the border.' [Pause] But I, I shouldn't complain, because within four years the longest time I spent at the border was 6 hours. And when I talk with my friends … they sometimes had to spend 17 or 20-something hours there. So, 6 hours compared to 20 is nothing.'

This account draws attention to the fact that the existence of a border, time consuming border crossing and border restrictions makes physically close places symbolically very distant. In contrast to integration processes taking place within internal Schengen border regions, in youth narratives the Polish–Ukrainian border precludes the closeness effect. Prompted on their visions of the border and in response to the hostile and institutionalized border environment some students develop stories of a dream border – an open border, without lines and check procedures; a border that welcomes instead of keeping away, and a border which would be similar to one of those EU internal boundaries. For example, one of the Polish students in Lviv (a regular cross-border commuter) comments:

> T4: 'I would really like to see this border checkpoint without a car line one day. Like between Poland and Slovakia. That would be really great.'

Another student, asked to depict the border, divided her drawing space into two and based on opposites sketched a 'realistic' border and a 'dream' border. The realistic border illustrates the existing problem of a hard border – border regime tightening, control done by border/duty officers and uncertainty about the possibilities of entering the other side. The dream border – the future of the border region as the student silently suggests – illustrates only an empty road with a 'Welcome in Ukraine' sign just passed by. The empty road as a symbolic ideal border reproduces the image of many present EU internal borders. Perceiving the Polish–Ukrainian border region through the prism of a typical EU reality is one of the examples of the significant impact the European Union may have on youth identity.

The specific Euro-curtain (Bialasiewicz and O'Loughlin, 2002) – 'paper' administrative firewall (Walters, 2006b) – protects the European community from outside influences. It is an invisible implication of the Schengen border regime and has a consequence in social construction of otherness and unfamiliarity, symbolic division into EU citizens and non-EU inhabitants. This perspective is ignored by the Polish interviewees. For Poles, the act of crossing the Polish–Ukrainian border differs from the fast and friendly borders within the Schengen zone only very briefly. Specifically, it requires certain organizational difficulties and formalities. But, no further proceedings are required. Therefore, it is not surprising that Polish interviewees hardly ever refer to the hard border consequences. Ukrainian students, on the other side, notice the border problem and supply with various stories of better–worse relationship, stigmatization and otherness, group of the privileged and administrative isolation followed by symbolic and mental cut-off from Europe.

A female Ukrainian student from the Polish minority reveals how deeply rooted such self-perceptions may be. Her mother is a Pole and her father, a Ukrainian. She was brought up in Ukraine in a mixed environment, and claims to 'feel' neither entirely Polish, nor Ukrainian. In the border discussion, however, she clearly sympathizes with Ukrainians (which she identifies with by using the grammar form: *we*). The symbolic exclusion of Ukraine from the European club results in a feeling of being worse and helpless. She stresses both nations' historical and geographical closeness and her difficulty in understanding why one of the parties is more restricted in its rights than the other. In her narrative the very fact of being outside the EU implies the worse international status of the country. The EU itself is perceived as a pass to a better world where one can literally 'rock'.

> T15: '(The border) it's an annoying place. I'd like it not to exist. It would be so nice. ... Yeah. When you enter Poland you can *rock*! [Pause] You know, one has this inferiority complex, because Poles can move and travel wherever they want. How often they want. And they are our neighbours we have a common history ... and suddenly it turns out we're somehow worse. Those actions of enforcing the eastern EU border – it's kind of sad, one feels worse.'

It seems that a direct crash with the administration, the formalities and the law result in a sense of isolation and exclusion. Another Ukrainian interviewee speaks of 'evident discrimination' implied by the border existence in the institutionalized form. Taking the foot border checkpoint he can see that the EU-members gate line is much longer, because the EU citizens may easily cross the border without any earlier preparations. And, apparently, they (mainly Polish people) prefer to stay in the EU-citizens line, which visually confirms their belonging to the Community, than take All-passports gate, which is considered the one for non-EU inhabitants. But the line in the All-passports gate is much smaller because of the other simple reason – very few Ukrainians can afford and/or obtain documents to travel to Poland.

> T16: 'The discrimination at this border is very evident. There are two separate lines: one for people from the EU, and the other for people from non-EU countries. And when you cross the border to get to Poland you see that there are 20 people in this EU-citizens line and only 2 in All-passports, in this case Ukraine, line – it's sad.'

Interestingly enough, a Ukrainian Ph.D. student in Lublin considers the whole border region an entity which has been forcefully divided by the border. She is the one to name the division an iron curtain isolating Ukraine from Europe. The interviewee admits that – living in Poland herself – she feels trapped because her contact with her family and Ukraine is restricted by the existing EU-imposed law. In her story, the visa regime is not necessarily discriminatory but first and foremost absurd and illogical. The interviewee recollects her travel difficulties and contacts with the embassy officers. The regulations make her feel helpless and entirely dependent on the system she cannot comprehend. Her struggle with the invisible bureaucracy may remind stories of an individual trapped in a totalitarian system.

The one facing the system feels 'like a third rate kind of person' just because he or she originates from the outside.

> T13: 'The visa system is very strange – even though Poland is already within Schengen (zone), what we get is only a national visa, which means – I can only be in Poland. If I wanted to go to ski in Slovakia or to visit somebody in Germany – I cannot do it. For me – it's a tragedy. Last year, when I was doing an internship in Prague, I had to have Polish and Czech national visas. I also get a combined one: national and Schengen, but Schengen was only for 3 months … Nobody cares about one's plans – last year I had been planning to go to a ball to Vienna. I had had everything prepared. I went to Austrian embassy in Prague and told them I had to go for 3 days to Vienna. They answered they couldn't give me a visa, because previously I had already had a 3-month visa and another one I can obtain only after the next six months. I said: "I'm right now within Schengen territory (in Czech Republic)". "Yes, you are." "I can be in Poland – it's Schengen too. So, why I cannot be in Austria – also Schengen? Where's the logic?" [Laugh] And those officers there smiled and said: "That's the law – we cannot do anything. That's the law." So … it's the only thing that makes me feel uncomfortable. I feel like a third rate kind of person. "You can stay only here. Be happy with being here."'

In her narrative the new iron curtain around the Fortress Europe does not have a firm visual form – it is neither the wall, nor even the gate or fence of a border checkpoint. The new iron curtain is a paper curtain (Grabbe, 2000) built up from formalities, official writs and bans – visas, invitations the non-inhabitants need to possess, insurances, confirmations, fees and payments that have to be delivered, agreements and proceedings. In addition, it is a soft-silk curtain implying hours lost in the lines to consulates, hours lost during personal checks, product transfer restrictions and limitations, agreements only for temporary stays, the look the border officer allegedly give to the outsiders, upsetting why-where-when questions during the personal check, feeling of being excluded from civilization and the space one believes one originates from, human helplessness and involvement into bureaucracy, sense of isolation, ethnic stigmatization and dreams of empty roads connecting both countries... which altogether result in a developed belief of being 'worse'. The tragicomic status of the border and the iron curtain effect are very accurately commented in a seemingly funny metaphor provided by the Ukrainian student:

> T16: 'You know, I have this association – it's a funny comparison … that the border (Polish–Ukrainian) is like an ECCO sole – it lets the air out, but doesn't let the moisture in. The same here – when somebody from the West wants to come to Ukraine it's – "be welcome". But if you yourself (a Ukrainian) want to go somewhere – it's a problem. Tell me, why this problem exists only for the people from the eastern side of the border?'

The process of hardening the external EU border takes a form of an administrative policy which non-EU students do find discriminatory. The air in the interviewee's story relates to the EU citizens being able to cross the border easily and comfortably. The moisture on the other hand indicates something undesirable which should be stopped or avoided by all means. This analogy expresses the essence of contemporary Fortress Europe – a gated community protected by a firewalling policy of letting in those who are acceptable and desirable while keeping away those who do not fit the profile. The policy has its obvious visible dimension (border checkpoints). But it is the invisible administrative firewall that in the end constitutes the privileged and fortified zone.

Schengen(less) Stories

Paradoxically, although being a popular theme in mass media and institutionalized EU discourse, Schengen as a labelled phenomenon does not appear in youth narratives very often. Interestingly, in the light of rich perceptions mentioned, the border is rarely associated with the Schengen border and the name usually appears no sooner than after being suggested by the interviewer. Then, it is most often connected directly with border crossing practices: restrictions, goods limitations, personal check and a visa.

For the majority of the interviewed Polish students Schengen as a term does not imply any symbolism or significance. In their narratives it becomes devoid of a deeper socially constructed sense. In other words, there is no conceptual recognition; Schengen is only a buzzword which fuels rough discomfort of facing border procedures. It seems that they develop a relatively indifferent and ideologically-free attitude towards the border as the Schengen border. In contrast to the notion of 'Polish–Ukrainian border' which arouses strong emotions, the 'Schengen border' is considered a social and political fact which does not become a subject of evaluation, criticism or questioning.

On the other hand, Ukrainian interviewees much more often provide a deeper meaning for Schengen. A Ukrainian student in Lublin seems to associate the term with the discriminatory dimension of the EU external border.

> T14: 'People probably feel like sufferers a little … Of course, they feel worse, because they have problems with getting a visa … because there are huge lines to the consulates.'

The student's narrative reflects emotional engagement – empathy and compassion. Similar attitude to Schengen as a notion is represented by one Polish young woman. The voice, though, is clearly isolated from those of the other Polish interviewees. In reaction to the Schengen topic, the interviewee stresses that the new border regime was imposed without a reason, just as a consequence of Poland's membership of the European Union. The speaker considers it discriminatory for everybody who is excluded from the club of 'the privileged'. The unprivileged, the worse, non-EU inhabitants – the interviewee seems to say – on the top of hardly acceptable financial situation, are forced to apply for costly visas.

> T5: 'Maybe Ukrainians … they may feel discriminated a little. Because suddenly – well, not suddenly – but without a reason … Why are they supposed to be worse? Why shouldn't they have a privilege we have?'
> I: 'You used an expression "they're worse" … '
> T5: 'Yes. Because a visa, for example, costs X euro and only very few people can afford it. So, in this material [financial – authors' remark] respect, they feel, I think, worse … '

A very interesting opinion in the Schengen impact discussion is also provided by one of the interviewees being member of the Polish minority in Ukraine. The interviewee mentions the administrative changes which financially affect many Ukrainians. But below the – in a sense superficial – discriminatory level, she also seems to notice something more than the material aspect of the contemporary Polish–Ukrainian neighbourhood. In her account Ukrainians from the direct borderland does not perceive themselves in relation to Schengen, the EU or any post-national administration. The Schengen exists only on a formal (border-crossing) level. The region stays a two-nation neighbourhood and the border discourse does not exceed the national identity level.

> T8: 'Many people are in a worse situation right now. It's difficult for them to come to the other side. Visas, papers, limits, time restrictions. The rules changed … People used to earn money selling things on Polish side and right now … it stopped … But [pause] generally, people don't consider themselves some "far East" bordering Schengen. In reality, in their minds, Schengen border doesn't mean anything. Like if it didn't exist. Well, it exists when those how-much-cigarettes-I-can-take issues are considered.'

In general, Schengen recognition is most often superficial and hardly ever relates to phenomena other than practicalities – border checkpoint procedures or visas. In relation to those administrative rules mainly Ukrainian interviewees find the Schengen-zone a privileged space which implies the worse position for those who come from the outside.

Discussion

Both Poland and Ukraine have experiences with being on the fringe of Europe defined as the EU (Wolczuk, 2009). However, at present, these two countries stay in a completely different relation to the EU. Poland has been benefiting from the EU membership since its accession, while Ukraine's status as a possible EU candidate is still contested. Both countries, having a long common history especially with reference to the post-Galician region, have been facing a 'new' state of the same border since 2004. During the first years its significance was more symbolic; now, from late 2007 on – with the Schengen regulation fully introduced – it has also been the deeply protected border.

The Polish–Ukrainian border as an example of the EU external boundary has not been addressed in youth studies yet. In the chapter, we have analysed the border stories that a sample of young migrants develop as a consequence of a regular border crossing and organization of everyday life in two neighbouring countries. In their narratives, it has been shown that the Polish–Ukrainian border is perceived twofold: directly, through the prism of everyday checkpoint experiences and indirectly, as a metaphor of the widely understood border regime imposed by Poland's EU accession. The checkpoint becomes an embodiment of the border meant classically as a traditional hard border between states. However, the youth approaches the concept referring to their positive experiences with or socially transmitted images of the Schengen internal borders. Therefore the border as checkpoint assumes the proportions of a total institution with a highly specific informal culture developed. Metaphorically, the border plays a role of a distancing and isolating factor, a bureaucratic, far from a dream construction established by means of the supranational protectionist policy.

The analysis of the border narratives derives its conceptual background from the criticized, although in the light of the recent EU enlargements still relevant, Fortress Europe metaphor as well as most contemporary gated community and firewall analogies. It has been argued that the (in)visible wall of Fortress Europe do exist in youth stories. It takes various forms and relates to different discoursive levels since bordering, othering and separation are perceived extremely distinctively by humans. It is often ephemeral and quite different from the classic Fortress Europe concept we might be acknowledged with. Although being relatively rarely directly labelled or addressed in the narratives, the wall clearly emerges out of the hard border regime the youth faces, the recollected checkpoint struggles, bureaucratic procedures and personal feelings. In the same time this wall stays surprisingly underidentified, fuzzy and often detached form the Schengen context.

Interestingly, the Schengen border recognition is particularily weak among the young EU insiders who find it no more than mere rhetoric. Mainly in the Ukrainian stories Schengen becomes a meaningful notion embracing socially constructed significance. However, in cases of both – Polish and Ukrainian students, it remains a figure of speech diverged from its original meaning, framed and spread as a word of mouth without any further reflection on what the Schengen Agreement or Schengen regime in reality are.

What strikes is the fact that the majority of border stories are narratives of exclusion. They are seldom devoid of social, historical and political significance. There are interviewees who explicitly reproduce and enrich with new meanings the Cold War era relict – the Iron Curtain. This issue proves worthy of a further investigation. Specifically, it shall be explored whether perceptions of the Polish–

Ukrainian border and other post-Cold-War boundaries as borders of exclusion are more widely or differently represented in youth narratives.

The article strives to grasp the changes in the social construction of the Polish–Ukrainian border which followed the Schengen zone Eastern enlargement. Although, it focuses mainly on the status of the particular border within the particular region, we hope it could become an exemplary study of the influence the strengthening of the EU border has for the non-EU members and the 'Schengen border regions'. Simultaneously, we believe the research contributes to a bigger discussion about the construction of youth border identities. Undoubtedly, further investigation needs to be done within this and the other EU borderlands to explore diverse (but hypothetically similar) effects the external border might have for the regions themselves as well as the image of the EU as an institution which not only contributes to disappearance of boundaries but also to the establishment of the new.

References

Armbruster, H. and Meinhoff, U.H. (2005), 'Memories of Home? Narratives of Readjustment on the German/Polish and former German/German Borders', in J. Coates and J. Thornborrow (eds), *The Sociolinguistics of Narrative*, Amsterdam: J. Benjamins Pub. Co., pp. 41-65.

Armbruster, H., Rollo, C. and Meinhof, U.H. (2003), 'Imagining Europe: Everyday Narratives in European Border Communities', *Journal of Ethnic and Migration Studies* 29(5): 885–99.

Berdahl, D. (1999), *Where The World Ended: Re-unification and Identity in the German Borderland*, Berkeley: University of California Press.

Berg, E. and Ehin, P. (2006), 'What Kind of Border Regime is in the Making? Towards a Differentiated and Uneven Border Strategy', *Cooperation and Conflict* 41(1): 53–71.

Bialasiewicz, L. and O'Loughlin, J. (2002), 'Galician Identities and Political Cartographies on the Polish–Ukrainian Border', in D.H. Kaplan and J. Häkli (eds) *Boundaries and Place: European Borderlands in Geographical Context,* Lanham, MD: Rowman and Littlefield, pp. 217–38.

Boedeltje, F.J., van Houtum, H. and Kramsch, O.T. (2006), 'Communication from the European Union: The European Neighbourhood Policy (ENP) "The Latest Addition to our Democratization Toolbox"', 'EUDIMENSION' STATE OF THE ART STRATEGY PAPER Nijmegen.

Castells, M. (2000), *The Rise of the Network Society*, Oxford: Blackwell Publishers.

Chojnowski, A. and J.J. Bruski (2006), *Ukraina*, Warszawa: Wydawnictwo Trio.

Copsey, N. (2008), 'Remembrance of Things Past: the Lingering Impact of History on Contemporary Polish–Ukrainian Relations', *Europe-Asia Studies* 60(4): 531–60.

Creswell, J.W. (2007), *Qualitative Inquiry and Research Design: Choosing Among Five Approaches*, Thousand Oaks, CA: Sage Publications.

Davies, N. (2008), *Boze igrzysko: Historia Polski*, Kraków: Wydawnictwo Znak.

Dekker, H. and Jansen, L.B. (1994), 'In Search for an Explanation Of Anti-German Attitudes Among Dutch School Children', in G. Csepeli (ed.), *From Subject to Citizen*, Budapest: Hungarian Academy of Sciences, Hungarian Centre of Political Education, pp. 105–22.

Dekker, H. and Jansen, L.B. (1995), 'Attitudes and Stereotypes of Young People in the Netherlands with Respect to Germany', in S. Hübner-Funk, L. Chisholm, M. du Bois-Reymond and B. Sellin (eds), *The Puzzle of Integration: European Yearbook on Youth Policy and Youth Research, vol. 1*. Berlin, New York: Walter de Gruyter, pp. 49–61.

du Bois-Reymond, M., Dekker H. and Aspelach R. (1998), 'Attitudes of Dutch Young People towards Germany and the Germans', in CYRCE, M. du Bois-Reymond and S. Hübner-Funk (eds), *Intercultural Reconstruction: Trends and Challenges*, Berlin: Walter de Gruyter, pp. 99–116.

Earthy, S. and Cronin A. (2008), 'Narrative Analysis', in Nigel Gilbert (ed.), *Researching Social Life*, London: Sage Publications, pp. 420-39.

Gänzle, S. (2009), 'EU Governance and the European Neighbourhood Policy: A Framework for Analysis', *Europe-Asia Studies* 61(10): 1715–34.

Goffman, E. (1961), *Asylums: Essays on the Social Situation of Mental Patients and other Inmates*, New York: Doubleday.

Grabbe, H. (2000), 'The Sharp Edges of Europe: Extending Schengen Eastwards', *International Affairs (Royal Institute of International Affairs 1944–)* 76(3): 519–36.

Haase, A. (2005), 'Between Integration and Exclusion: Ethnic Minorities at the Polish Eastern Border after the First Decade of Transition', *National Identities* 7(2): 213–30.

Haukkala, H. (2008), 'The European Union as a Regional Normative Hegemon: The Case of European Neighbourhood Policy', *Europe-Asia Studies* 60(9): 1601–22.

Hrytsak, Y. (2005), 'The Borders of Europe – Seen from the Outside Eurozine', http://www.eurozine.com/articles/2005-01-10-hrytsak-en.html (accessed 22.8.2010).

Iglicka, K. and Weinar, A. (2008), 'Ukrainian Migration in Poland from the Perspective of Polish Policies and Systems' Theory', *Journal of Immigrant & Refugee Studies* 6(3): 356–65.

Jukarainen, P. (2003), 'Definitely not yet the End of Nations: Northern Borderlands Youth in Defence of National Identities', *Young* 11(3): 217–34.

Kasianov, G. (2006), 'The Burden of the Past: The Ukrainian–Polish Conflict of 1943/44 in Contemporary Public, Academic and Political Debates in Ukraine and Poland', *Innovation: The European Journal of Social Science Research* 19(3): 247–59.

Klatt, M. (2011), 'Poland and its Eastern Neighbours: Foreign Policy Principles', *Journal of Contemporary European Research* 7(1): 1–16.

Klein, N. (2003), 'Fortress Continents', *The Guardian*, 16 January.

Kofman, E. and Sales R. (1992), 'Towards Fortress Europe', *Womens Studies International Forum* 15(1): 29–39.

Kordan, B. (1997), 'Making Borders Stick: Population Transfer and Resettlement in the Trans-Curzon Territories, 1944–1949' *International Migration Review* 31(3): 704–20.

Kvale, S. and Brinkmann, S. (2009), *InterViews: Learning the Craft of Qualitative Research Interviewing*, Los Angeles: Sage Publications.

Maxwell, J.A. (2005), *Qualitative Research Design: An Interactive Approach*, Thousand Oaks, CA: Sage Publications.

Meinhof, U.H. (ed.), (2002), *Living (with) Borders: Identity Dscourses on East–West Borders in Europe*, Aldershot: Ashgate.

Meinhof, U.H. (2003a), 'Migrating Borders: An Introduction to European Identity Construction in Process', *Journal of Ethnic and Migration Studies* 29(5): 781–96.

Meinhof, U.H. (2003b), 'Migrating borders: an introduction to European identity construction in process', *Journal of Ethnic and Migration Studies* 29 (5): 781–796.

Meinhof, U.H. and Galasinski, D. (2000a), *Border Discourse: Changing Identities, Changing Nations, Changing Stories in European Border Communities: A state-of-the-art report,* The European Border Identities Consortium.

Meinhof, U.H. and Galasinski, D. (2000b), *Changing Identities, Changing Nations, Changing Stories in European Border Communities: A 'state-of-the-art' report by Ulrike Hanna Meinhof and Dariusz Galasinski in collaboration with the European Border Identities consortium of the Framework V project No. SERD-1999-00023.*:

Meinhof, U.H. and Galasinski, D. (2000c), 'Photography, Memory, and the Construction of Identities on the Former East–West German Border', *Discourse Studies: An Interdisciplinary Journal for the Study of Text and Talk* 2(3): 323–53.

Meinhof, U.H. and Galasinski, D. (2002), 'Reconfiguring East West identities: Cross-generational Discourses in German and Polish Border Communities', *Journal of Ethnic and Migration Studies* 28(1): 63–82.

Meinhof, U.H. and Galasinski, D. (2005), *The Language of Belonging*, Basingstoke: Palgrave Macmillan.

Mrinska, O. (2006), 'The Impact of EU Enlargement on the External and Internal Borders of the New Neighbours: The Case of Ukraine', in J.W. Scott (ed.), *EU Enlargement, Region Building and Shifting Borders of Inclusion and Exclusion*, Aldershot: Ashgate, pp. 81–93.

Orbie, J. (ed.), (2008), *Europe's Global Role: External Policies of the European Union*, Aldershot: Ashgate.

Paasi, A. (1998), 'Boundaries as Social Processes: Territoriality in the World of Flows', *Geopolitics* 3(1): 69–88.

Petrat, A., Kindervater, A., Rippl, S. and Boehnke, K. (2009), *Transnationale Zivilgesellschaft: Einsichten under Perspektiven aus Städten in Grenzregionen. Europa im Wandel*, Frankfurt am Main: Peter Lang.

Rumford, C. (2008), 'Introduction: Citizens and Borderwork in Europe', *Space and Polity* 12(1): 1–12.

Sagan, I. (2010), 'Post-Socialist Transformation, European Neighbourhood and Civil Society Networks between Poland, Russia and Ukraine: a Case of Multi-level Contingency', *Journal of European Integration* 32(5): 439–56.

Sahlins, P. (1989), *Boundaries: the making of France and Spain in the Pyrenees*, Berkeley: University of California Press.

Sahlins, P. (1998), 'State formation and national identity in the Catalan', in T.M. Wilson and H. Donnan (eds), *Border Identities: Nation and State at International Frontiers*, Cambridge: Cambridge University Press, pp. 31–61.

Sandberg, M. (2009), 'Grænsens nærvær og fravær: Europæiseringsprocesser i en tvillingeby på den polsk-tyske grænse', Ph.D-thesis, University of Copenhagen, Section for Ethnology at the Saxo-Institute.

Schultz, H., Stoklosa K. and Jajesniak-Quast, D. (2002), *Twin Towns on the Border as Laboratories of European Integration*, Frankfurt an der Oder: Frankfurter Institut für Transformationsstudien.

Scott, J.W. and Matzeit S. (eds) (2006), *EXLINEA Lines of Exclusion as Arenas of Co-operation: Reconfiguring the External Boundaries of Europe – Policies, Practices, Perceptions, FINAL PROJECT REPORT*.

Snyder, T. (2005), 'The wall around the West Eurozine', http://www.eurozine.com/articles/2005-01-06-snyder-en.html (accessed 16.8.2010).

Stoklosa, K. (2003), *Grenzstädte in Ostmitteleuropa. Guben-Gubin 1945–1995*, Berlin: Berliner Wissenschaftsverlag.

Szmagalska-Follis, K. (2009), 'Are the European Union's New Boundaries like the Iron Curtain? 1989, Borders and Freedom of Movement in Poland and Ukraine', *International Journal of Politics, Culture, and Society* 22(3): 385–400.

Van Houtum, H. and Pijpers, R. (2005), 'Towards a gated community', Eurozine, http://www.eurozine.com/articles/2005-01-12-houtumpijpers-en.html (accessed 18.8.2010).

Van Houtum, H. and Pijpers R. (2007), 'The European Union as a Gated Community: The Two-faced Border and Immigration Regime of the EU', *Antipode* 39(2): 291–309.

Walters, W (2006a), 'Border/Control', *European Journal of Social Theory* 9(2): 187–203.

Walters, W. (2006b), 'Rethinking Borders Beyond the State', *Comparative European Politics* 4(2–3): 141–59.

Wolczuk, K. (2002), 'The Polish–Ukrainian border: On the Receiving End of EU Enlargement', *Perspectives on European Politics and Society* 3(2): 245–70.

Wolczuk, K. (2009), 'Implementation without Coordination: The Impact of EU Conditionality on Ukraine under the European Neighbourhood Policy', *Europe-Asia Studies* 61(2): 187–211.

Yekelchyk, S. (2007), *Ukraine: Birth of a Modern Nation*, Oxford/New York: Oxford University Press.

Ziolkowski, M. (2008), *Projekt: Ukraina*, Wroclaw: Kolegium Europy Wschodniej im. Jana Nowaka-Jezioranskiego.

Chapter 10
Living (beyond) the Border: European Integration Processes and Cross-border Residential Mobility in the Italian–Slovenian Border Area

Devan Jagodic

Introduction

The idea that the last decades in the European Union have born witness to a certain degree of state power erosion and that the borders between states are getting more and more permeable seems to be gathering consensus (Blatter, 2001; Perkmann and Sum, 2002; Bufon, 2006). Even if stating that the nation-states are an endangered species would be excessive, mobility across borders is currently an everyday practice for millions of European citizens. The Schengen Agreement, signed by 28 states, has progressively cancelled customs controls, thus allowing people to move freely from one state to another. These developments seem to be particularly relevant for the EU internal border areas. The 'open border' enhances mobility and multiplies the opportunities of meeting and cooperating with bordering economies and cultures. The intensity of such cross-border relations emerges from the spatial mobility of citizens, who can satisfy their needs linked to supplies, work, education, services, free time, tourism, etc. These forms of mobility, which contemplate frequent movements between the origin and the destination country, are destined to question the concept of national borders.

The chapter provides insight into one of the emerging phenomena of spatial mobility in EU internal border areas, for which the term 'cross-border residential mobility' (CBRM) is suggested. The process, which can be observed in several European border settings, refers to the migration movements produced by 'borderlanders', who decide to buy a house and move to the other side of the border, at the same time maintaining strong functional and affective relations with their previous country. Even if CBRM flows have their origins in the asymmetries between the bordering areas in terms of real estate prices and cost of living, they rapidly increased especially after the strong acceleration of the European integration processes during the 80s and the 90s (van Houtum and Gielis, 2006; Terlouw, 2008). The provisions of the Community, adopted in order to meet the goal of creating a common market, have facilitated the process of purchasing real estate in all of the EU territory, thus offering people the opportunity to access the real estate market in the neighbouring state or region.

Moreover, in a border area with a stormy past as is that between Italy and Slovenia the CBRM is a phenomenon witnessing profound geopolitical and socio-economic changes. The processes of European integration and regionalization have undoubtedly contributed to the spread of an atmosphere of greater relaxation in the relations between the two neighbouring states, while at the same time easing cross-border circulation. The entrance of Slovenia into the EU in 2004 and the consequent liberalization of its real estate market have created a growing Italian demand for housing in the Slovenian rural areas of the nearby Karst. Indeed, the Karst territory not only provides a high residential quality due to the presence of vast green areas, but also offers favourable real estate prices in comparison to Italian standards. The opportunity to improve living conditions by moving just a couple of kilometres further, in an area that can be reached with no obstacle whatsoever, has therefore encouraged – and is still encouraging – many Italian citizens to make this kind of decision.

The chapter is divided into three parts. The first part will present a review of the (few) academic papers having dealt with the phenomenon of CBRM so far. The comparison of four European case-studies will enable the location of the main causes, features and development paths of these processes. The second part is an attempt to provide the studied phenomenon a proper conceptual framework by approaching it within two analytical perspectives: suburbanization theories and transnational migration studies. The third part will draw attention to the Italian–Slovenian borderland. The information obtained from a recent study based on quantitative data will allow for the detection of congruencies with previously presented case-studies and the suggested theoretical framework.

CBRM in the EU Internal Border Areas: Analysis of Four Case-studies

The CBRM has not yet received thorough academic attention. To the best of the author's knowledge, only a few scientific papers have dealt with such relatively new ways of crossing the national borders. In the EU scenario, the case-studies that have been studied so far regard the border areas between the Netherlands and Germany (Strüver, 2005; van Houtum and Gielis, 2006), the Netherlands and Belgium (van Houtum and Gielis, 2006), Germany and France (Terlouw, 2008), and Slovakia and Hungary (Hardi, 2009; Lampl 2010). Other studies have examined the influence of the real estate market in determining cross-border commuting; for instance in the borderlands between Switzerland and France (Bolzman and Vial, 2007) or Germany and Denmark (Hansen and Schack, 1997). Information and statistical data on the Danish–Swedish case can be found on the Internet (TendensØresund, 2011).

In the following pages, four case-studies corresponding to four EU internal border areas that have recently experienced a constant growth of cross-border residential mobility flow will be examined. The review includes three borderlands between the EU-15 states (Netherlands–Germany, Germany–France, Denmark–Sweden), where belonging to the 'European common home' is a rather consolidated fact, and a borderland between the EU-12 states (Slovakia–Hungary), which entered the Union in 2004 and where the physical borders were abolished only in 2007.

The Dutch–German Borderland

The first case-study covers the area between the Netherlands and Germany, one of those European borderlands that in the last two decades experienced the loss of their constitutive property: the border. The abolition of the physical barriers thorough the Schengen agreement in 1985 has progressively modified the attitude of the population towards the border, which is no longer perceived so much as a barrier, but a source of attraction and opportunities (Spierings and van der Velde, 2008). Among the new practices of crossing the Dutch–German border is the growing CBRM flows by Dutch citizens, who have decided to move to the neighbouring German areas. Beyond the new psychosocial climate, promoted by European integration processes, the economic motivations seem to have had a crucial role in these decisions. The main 'push factors' (conditions that drive people to leave their homes) can be identified in the lack of available houses and the high prices on the Dutch side, while the 'pull factors' (conditions that attract people to the new area across the border) concern the wide availability of houses and lower prices on the German side, along with the cheaper cost of living in Germany (van Houtum and Gielis, 2006).

The Dutch migrants mostly come from an urban environment and are attracted by the high quality of life offered by the neighbouring German rural areas (Strüver, 2005). They tend to settle in residential complexes a few kilometres from the border line, which were built *ad hoc* to accommodate the high Dutch demand. These agglomerations can be considered as satellite settlements of the Dutch border cities, where the urban populations find favourable residential conditions. Some examples of these settlements are Bunde (near Winschoten), Bad Bentheim and Gronau (near Enschede), Selfkant (near Sittard), and most of all Kranenburg, a German rural municipality a few kilometres from Nijmegen, a Dutch border city with 170,000 inhabitants.

In the last twenty years, the rural landscape of Kranenburg, which is typical of the villages lying along the Rhine, has experienced deep changes by acquiring more and more urban elements. These landscape transformations are a clear consequence of the recent massive incoming of Dutch urban population and of the residential buildings constructed to host it. Indeed, by the year 2005 the Dutch citizens having moved to Kranenburg have already reached 18.2 per cent of the whole population of the area (Huijgen and Reijmer, 2005).

Strüver (2005) studied the peculiar way of life undertaken by the Dutch community of cross-border migrants settled in Kranenburg. The data show that they tend to maintain strong bonds with the country of origin and especially with

Nijmegen, the city where they work, bring their children to school, visit friends and relatives, and spend their free time. Van Houtum and Gielis (2008) describe such phenomena as 'elastic' migration. Hence, the elastic is produced by the constant interaction between centrifugal forces (moving house in the new country) and centripetal forces (maintaining of functional and/or affective bonds with the country of origin).

The migrants' 'elastic' way of life and their apparent lack of desire to integrate into the new socio-cultural context generate a wide debate in the public opinion on the German side. Some media explicitly speak about the Dutch 'invasion' and label the new settlements as 'enclaves', 'colonies' or 'little Amsterdam'. The greatest objections regard two points: 1) the unidirectionality of the migration flows, which reflects the greater purchasing power of the Dutch population; 2) the rise in real estate prices caused by the Dutch demand, which tend to preclude German citizens from the opportunity to buy houses in their own territory.

The German–French Borderland

The second case-study refers to the German–French borderland in the Rhine Valley, founded by the PAMINA Euroregion in 1989. In this area, the relations between the two sides of the border do not just develop in the institutional framework of cross-border cooperation, but also include a series of 'cross-border behaviours' adopted by the population (Terlouw, 2008). These behaviours are mainly a consequence of the PAMINA's asymmetrical regional structure. National differences in economic prices and regulations make it profitable for many borderlanders to use both sides of the border in their daily lives. These asymmetries produce processes of cross-border commuting, which takes course in two ways. The first is represented by the French border workers, who live in Alsace and work in Germany, the second by the German citizens coming from Karlsruhe, who have moved to the French side while at the same time retaining their jobs and social networks in Germany.

CBRM does not represent an absolute novelty in PAMINA. For decades a limited number of Germans migrated to France, attracted by the traditional old Alsatian farms in the hills and the French way of life. Nevertheless, the Germans entered the housing market in the French border zone *en masse* after the introduction of the common market in 1992, which made buying a house across the border easier. This enabled the Germans to take full advantage of lower French real estate prices. Comparable houses in Germany are several times more expensive, while tax levels are tens of per cent higher (Bohn, 1997). In some villages, the German migrants comprise 35 per cent of the total population.

These massive settlements have been encouraged by the generous building plots granted by the local municipalities. Some Alsatian villages with large new housing estates, such as Wintzenbach or Beinheim, register a high presence of German citizens, who live concentrated in separate new neighbourhoods, sometimes depicted as ghettos (Ramm, 1999). Language is a crucial factor in the continuing migrants' orientation towards Germany. Two-thirds of the German citizens living across the national border do not speak French (Bohn, 1997), while the willingness of the Alsatians to speak German in the social context of their own village is questionable. Some Germans learn French and integrate socially, but especially for many of the modestly educated, the language barrier becomes higher over time, so that, many decide to return to Germany (Terlouw, 2008).

The Danish–Swedish Borderland

The main feature of the Danish–Swedish border area is the presence of a natural barrier, the Øresund strait, which separates the regions Sjælland (Denmark) and Skåne (Sweden). In 2000, the two sides of the strait were joined by the construction of the Øresund Bridge, which greatly increased the volume of mobility flows between the two countries. From that moment onwards we witness a constant rise in the amount of cross-border commuters living in Skåne and working in København, attracted by the favourable working conditions offered by the regional capital city. Moreover, in the last decade cross-border commuting does not include only the Swedish, but to increasing extent Danish citizens who decide to transfer their residence to the other side of the strait.

CBRM is also not a novelty at the Danish–Swedish border. What distinguished this specific borderland from the others is that the proportion of the migration flows has drastically increased after the opening of the Øresund Bridge. Some recent researches clearly show how in the last ten years the number of Danish citizens moving to Sweden has tripled, passing from 1,582 transfers in 2001 to 4,284 in 2006 and even 4,311 in 2007 (TendensØresund 2011). The flows are settled in the proximity of the bridge, so as to shorten as much as possible the trip to the working place in København. Among the Danish having moved to Skåne in 2005 and 2006, for instance, more than one third purchased a house within five kilometres from Svågertorp (the closest railway station to the bridge) and more than one half live within ten kilometres from the bridge. The metropolitan area of Malmö represents the main destination of the migration flows.

The results of a survey commissioned by the Malmö municipality in 2006 and carried out among 2,000 Danish citizens having moved to the metropolitan area in the same year[1] show two orders of factors promoting the advancement of the migration flows, both converging in the economic sphere. The first is represented by the opportunity to purchase cheaper houses compared to the standards in the København metropolitan area, while the second factor regards the lower cost of life in Sweden. In the country of origin: only 39 per cent of respondents used to live in jointly owned or tenant owned houses, while after the move the percentage raised to 84 per cent. The general improvement of living conditions therefore entices the Danish citizens to move to the new country.

1 Wonderful Malmö. Sample survey for Danes who have moved to Malmö during 2006, Malmö City, City Hall, December 2007.

The Slovak–Hungarian Borderland

Despite the fact that the Slovak–Hungarian border cannot be considered a land free from tensions, the cross-border movements are by now part of a consolidated tradition. Already in the socialist period there were forms of mobility linked to work and shopping practices, which were enriched in the following centuries by a spreading cross-border use of institutions and services (education, health, etc.). When the two countries entered the EU and Schengen area, the traditional practices were followed by other types of cross-border mobility. For instance, more and more frequently Hungarian businesses move to the other side of the border (for the more business-friendly taxation system) and the relocation of residences – the movement of the inhabitants to the other side of the border – has begun as well.

Hardi (2009) considers the recent flows of CBRM at the Slovak–Hungarian border as an evolution phase of the suburbanization processes of the cities of Bratislava and *Košice*. Bratislava, a dynamically developing capital city has created a suburban zone of considerable size and rapid growth around itself. Due to the location of the city – close to the border – and the characteristics of the real estate market, this suburban zone has already penetrated into the Hungarian side and a further dynamic strengthening of this suburbanization process is expected. A total of 3 to 5 thousand Slovak citizens are estimated to have moved to the Hungarian border region as of the autumn of 2008.

A recent survey of the CBRM flows in the Slovak–Hungarian area (Lampl, 2010) posits the 'elastic way of life' adopted by the cross-border migrants. 80 per cent of them, indeed, have kept their jobs in Bratislava, while the great majority enrol their children in Slovak schools. In the new settlement context the Slovak migrants spend their free time mainly at home, showing little interest in creating new social relations with the local community. On the other hand, this lack of integration into the receiving society does not seem to cause great concern among the new immigrants. Among the main problems, underlined by the interviewees, there are other difficulties, such as those linked to the lack of shops and services close to the new residence, and the deficient public transport.

Theoretical Framework

Despite the presence of some peculiarities linked to the single reference context, it can be observed that these case-studies show a series of similarities. Bearing in mind these convergences in the following pages an attempt will be made to provide CBRM with a proper theoretical framework. In line with the guidelines introduced, the phenomenon will be approached in the frame of two different analytical perspectives: suburbanization theories and transnational migration studies.

CBRM as a Suburbanization Process

In each of the examples, CBRM is the result of centrifugal drives from the border cities towards the metropolitan suburbs. The border crossing of Nijmegen, Karlsruhe, København and Bratislava in the nearby rural outskirts is therefore a clear example of the advancement of suburbanization processes. Suburbanization is the expansion of the urban tissue beyond the cities' administrative borders, which in this specific case are also state borders. The neighbouring rural areas progressively become residential zones, suburbs of the cities, their satellite settlements.

Among the causes for the development of CBRM, the majority are typical of suburbanization processes: high residential density in the cities and scarce availability of housing solutions; cheaper houses in the rural peripheries compared to the standards in urban areas; evolution of lifestyles in favour of larger houses out of the city centre and close to green areas; improvement of the traffic connections and an increase in private motorization capacity. An important role in determining the dimensions and directions of the phenomenon is also played by local policies in terms of building permits, credit conditions, investments in the development of the peripheries, etc.

As a specific phase of the cities' 'cycle of life' (van den Berg, 1981), suburbanization experienced a great spread in Europe starting from the 1970s. In recent decades, the territories surrounding several European cities have faced radical changes. In border areas, suburbanization has often been hindered by physical barriers (customs controls), juridical barriers (normative restrictions to the purchase of real estate in the neighbouring country) or mental barriers (diffidence towards 'the Other' across the border), which prevented the cities from expansion in their hinterlands. The recent acceleration of European integration processes has contributed to the removal of those barriers and allowed the residential flows to penetrate to the other side.

As seen in the analysis of the case-studies, the suburbanization process is not a neutral phenomenon; it may cause a series of collateral effects, many of which can be observed in the border areas. The decentralization of the population in the new peripheral residential spaces imply, for instance, a great increase of private car commuting from home to the work place, with the consequent pollution problems. The high consumption of soil, due to the steady construction of new buildings, implies elevated costs linked to the necessity to guarantee the new settlements adequate infrastructures, reducing at the same time agriculture land. The rural landscape of the periphery gains an ever growing amount of urban elements and risks losing its identity – being gradually transformed into a non-place (Augè, 1992). As described by Sennet (1970), in the new residential suburbs the culture of defensiveness and privatization tends to spread, which inhibits the contacts and relations between old and new inhabitants. Moreover, the in-migration of the new middle class creates a process of rural gentrification, in which long-term residents are increasingly displaced by the upward spiralling of real estate prices.

Beyond this traditional suburban impact, the approach to CBRM shall also consider the deep consequences in the ethnic and linguistic composition of the receiving society. In border situations, which are often already multicultural and multilingual, the high inflow of citizens coming from the neighbouring country can break up the existing balances. The risk of emerging conflict situations rises especially where the new migrant communities cannot speak the local language and thus create ethnic enclaves, often detached from the rest of the population and are little inclined toward social integration.

CBRM as a Short-distance Transnational Migration

CBRM is a phenomenon characterized by a bi-dimensional flow, composed by one single 'permanent' migration (transfer of residence from one side of the border to the other) and a myriad of 'temporal' migrations (commuting to/from the mother country). Indeed, the migrants tend to maintain quotidian relations with the cities of origin, where they retain their jobs, bring children to school, visit friends and relatives, spend their free time, and use a series of services. Such migrations are therefore characterized by living 'in-between' two national societies.

In line with the theoretical guidelines introduced by Strüver (2005) and deepened by Gielis (2009), it is suggested here to approach the phenomenon of CBRM within the emerging analytical perspective in the international migrations studies: transnationalism. Transnationalism can be defined as 'the process by which migrants build social fields that link together their country of origin and their country of settlement' (Glick-Schiller et al., 1992, p. 1). Thanks to the reduction of the transport and communication costs a growing number of people have the opportunity to live a double life, speak two languages and have regular contacts through the national borders (Portes et al., 1999). Transmigrants tend to 'maintain' social, economic and cultural bonds with their country of origin, while at the same time 'construct' new ones in the country of destination, feeding them through commuting practices. In this case, the event does not involve the same separation trauma arising in definitive migrations and the choice made is not so crucial for the individual life (Ambrosini, 2008).

The migration flows developing in border areas differ from the contemporary, long-distance transnational migrations because of the geographical proximity between the country of origin and the country of destination. In this regard, Strüver (2005) speaks of 'short-distance transnationalism'. Where this continuous bouncing between here and there mainly has a virtual character in most cases of transmigration (Vertovec, 1999), the short-distance migrations in borderlands also displays a strong physical elasticity. Because of the proximity of the neighbouring country 'the commuting of the transmigrants is wholly real and tends to develop on a daily basis' (Gielis, 2009, p. 605). An overbalanced transnationalism towards the country of origin may imply a low motivation towards integration in the receiving society, which leads to the perception of short-distance transmigrants as an undesirable element.

The controversial relation between mobility – transnational practices and cultural/identity models has already paved the way for a rich theoretical contention. While for some authors the transmigrant groups represent the embryos of new cosmopolitan and post-national hybridisms (Soysal, 1994), characterized by a contextual and changing relation with identities, cultures and belongings, other authors consider these groups as new forms of nationalism, which are dangerous for the civil and democratic coexistence in the hosting societies (Huntington, 1996). In border areas, these interpretational contrasts seem to emerge in an even more radical way, as the CBRM flows rapidly twists the existing balances, generated by the long-standing presence of the border.

CBRM in the Italian–Slovenian Border Area

The CBRM in the Italian–Slovenian border area is a recent phenomenon, which has yet to capture academic attention. The only scientific papers dealing with this topic are limited to evaluating the environmental and landscape impact of the phenomenon (Lah, 2008; Prosen, 2008). The main source of information is therefore to be found in the press, although they are not always reliable because of the surface nature of investigation, as well as the evaluations often being influenced by ideological prejudices and the tendency to confer features to cross-border migrants that are not empirically validated. The present chapter aims at partly bridging this knowledge gap by reporting some data drawn from empirically observable sources, which will be introduced by a preliminary description of the reference context.

Past and Present Population Movements

The territory between Italy and Slovenia is a rather sensitive and historically controversial borderland. During the 20th century, the region experienced several variations of the border line with consequent changes in the belonging of vast territories to different states. Each variation was followed by changes in the political-institutional system, ethnic repressions and definitions of rigid borders, which strongly marked the relations between the neighbouring populations and created a series of migration movements. At the end of World War II, for instance, there was a long and painful exodus of Italian speaking citizens from Istria and Dalmatia, who migrated to Italy for economic and political reasons after these regions had been annexed to Yugoslavia.

The nonsense of the border, separating areas tightly linked in the past, and the dramatic demographic upheavals after World War II, represented the premises to transform the region into a crisis area. But this situation failed to evolve. On the contrary, the European integration processes, the regional movements, and the increasing cross-border cooperation render the socio-political climate in these territories progressively more relaxed, while the Italian–Yugoslav border became ever more permeable. During the 1960s there was a huge increase in cross-border mobility practices, primarily fed by economic interests, and later by tourism. From Yugoslavia many cross-border workers (bricklayers, gardeners, shop assistants, waiters, care workers) began commuting to/from Italy, attracted by better labour conditions (Zago, 2004). The democratization of Slovenia in the early 1990s and its recent entrance into the EU seem to have accelerated this practice, while giving birth to new spatial mobility phenomena.

The Appearance of CBRM

The CBRM in the Italian–Slovenian border area is a phenomenon that saw its rapid spread in 2004, immediately after Slovenia entered the EU. The event led to the liberalization of the Slovenian real estate market, which became accessible to the citizens of all EU member states and therefore to Italian citizens. For the inhabitants of Trieste, the capital city of the Friuli Venezia Giulia region lying next to the Slovenian border, real estate just across the border suddenly became very desirable. More and more Italian citizens decided to buy a house and move into the neighbouring rural villages of the Slovenian Karst[2], area that not only offers housing solutions with favourable economic conditions, but also provides for high residential quality because of the vast availability of green spaces. On the basis of public data, Slovenian real estate purchased by Italian citizens in the period from May 2004 to December 2010 totals 1,014 units, of which 398 are included in the territory of the Slovenian Karst (Tax Administration of the Republic of Slovenia, 2011).

The CBRM phenomenon is the advancement of the suburbanization of Trieste. This process began in the first years of the second post-war period when the villages of the closest hinterland were repopulated with new inhabitants coming from an urban environment. The presence of natural barriers to the east (the Adriatic Sea) and political-administrative barriers to the west (the national border with Yugoslavia and then with Slovenia) rendered the urban expansion impossible beyond the territory of the province. This provoked intense building in the whole part of the Karst territory belonging to Italy. The recent changes in the geo-political and socio-economic scenario produced by the European integration processes has therefore 'offered' to suburbanization the opportunity to cross the border and penetrate into the Slovenian side.

2 The Karst indicates the limestone borderline plateau region in southwestern Slovenia extending into northeastern Italy. The Slovenian Karst covers an area of 851 square metres and 27,270 inhabitants, including the municipalities of Sežana, Divača, Komen, Hrpelje-Kozina and the karstal zones of the municipalities of Koper/Capodistria and Miren-Kostanjevica (Velikonja 2007).

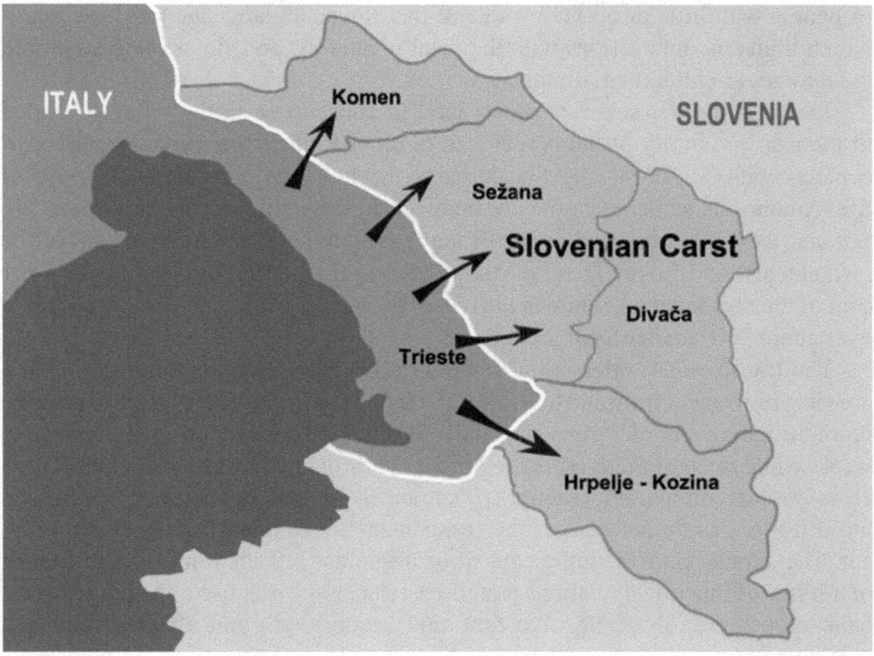

Figure 10.1 Suburbanisation of Trieste towards the Slovenian Karst

Transnational Experiences of the Cross-border Migrants

This section will briefly present the results from a survey carried out with a sample of Italian citizens having recently moved to the Slovenian Karst. The survey is part of the research project *Cross-border residential mobility in the context of EU: the case of the Italian–Slovenian border*, promoted by the Slovenian Research Institute – SLORI and developed by the author within his Ph.D. dissertationat the International University Institute for European Studies in Gorizia (Jagodić, 2009/2010). The data was obtained by a CAWI (Computer Assisted Web Interviewing) questionnaire, which was administered to 126 cross-border migrants. The respondents were identified by a snowball sampling technique, where a first group of cross-border migrants recruited other subjects with similar features from among their acquaintances.

According to the survey results, the actors of CBRM are predominantly young couples looking for their first house. The respondents were mostly working persons (88.9 per cent), many with salaried jobs. The interviewed sample presents a high level of education, as 36.8 per cent of the respondents held a university degree and 58.2 per cent high school graduates. 53.3 per cent of these young migrants are Slovenian speakers (most of them members of the Slovene national minority in Trieste), which is an element that inevitably eases the decision to move to the other side of the border. A greater problem is represented by the migration experience of people with little or no knowledge of the Slovenian language (39.1 per cent), which limits the interactions with the local population and slows integration into the new socio-cultural environment

The cross-border suburban flows tend to concentrate in the villages located in the near proximity of the border line in order to facilitate the maintenance of contacts with Trieste, a city that remains the migrants' gravitational centre of life. Among the settlements mostly involved in the inflow of Italian citizens are Sežana (the most urbanized centre of the region, targeted by 33.3 per cent of the respondents) and Lokev (a rural village next to the border, targeted by 10.6 per cent of the respondents), while in the rest of the territory the settlement is scattered over about 200 smaller local villages.

The transnational orientation of the cross-border migrants emerges in their constant movement from and towards Italy. Indeed, 66.7 per cent of the respondents cross the national border every day and a further 21.4 per cent do so 4–5 times per week, while none of them cross the border fewer than 2–3 times per week. These cross-border commuting practices illustrate a general dependency on private motorization, as 96 per cent of the respondents make these trips exclusively by car. The strong bonds with the context of origin are reflected in the maintenance of a series of functional relations (activities related to work, use of health services, bank procedures, shopping, free time and leisure, etc.) and affective relations (seeing/meeting relatives, seeing/meeting friends, evening/night outings, etc.) (Figure 10.2) For many of the interviewed subjects the new domicile is therefore used largely as a dormitory, a space to enjoy the 'country life' in the free hours, but still a space, which is perceived to have little to offer as regards infrastructures, services, as well as social and cultural life.

The overbalanced transnationalism towards the country of origin is further evident from the decisions regarding the instruction of the young generations. Indeed, among the interviewed families with children of pre-school and school-age there is a large preponderance of pupils enrolled in the schools in Italy, most of all in schools with Slovenian as the language of instruction[3] (45.5 per cent), but also in the schools with Italian as the language of instruction (21.2 per cent). Only a third of them (33.3 per cent) have opted for the enrolment of their children in the local Slovenian schools.

The centripetal orientation of the 'short-distance transmigrants' emerges as well in the widespread practice of maintaining the official residence in Italy (75 per cent of the respondents), as this choice implies the opportunity to retain open access to the national health care system. Some (31 per cent) even avoid communicating their presence to the Slovenian authorities, thus resulting to some degree in a 'hidden community'. Because of the lack of an integrated legal and

3 The schools of the Slovenian minority in Trieste are a constituent part of the Italian school system and dependent on the Italian Ministry of Public Education. The instruction is based on the same curriculum that is used in the schools with the Italian medium of instruction but takes place in the Slovenian language.

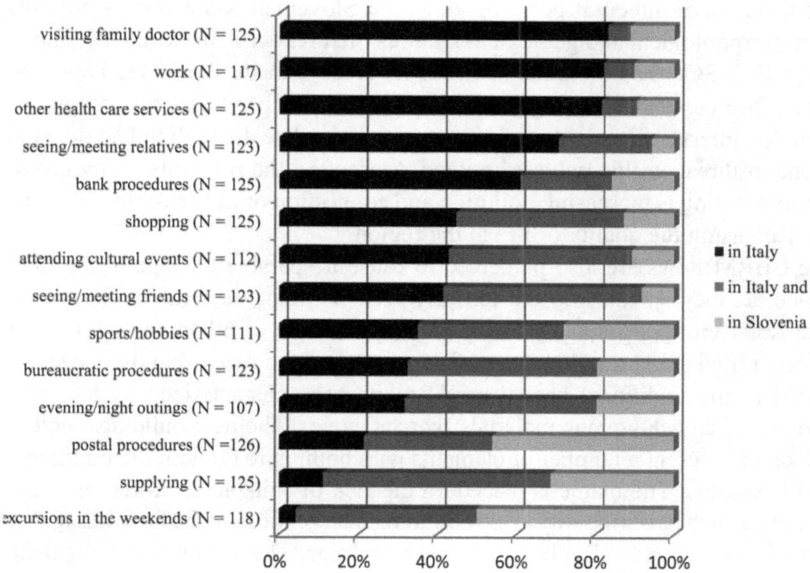

Figure 10.2 Transnational activities of Italian citizens having moved to Slovenian Karst

administrative framework, the cross-border migrants have a wide margin, within which they can choose between the two systems (Italian and Slovenian), depending on the personal benefits they can accrete. Strong ambiguities emerge in the taxation issue. Indeed, migrants generally pay their taxes to the Italian Treasury, as Italy is the country where they work and produce income. Thus a discriminatory situation against the Slovenian state arises, as the new residents use the Slovenian public infrastructure and services, while not contributing to their financing.

Reactions in the Slovenian Public Opinion

In a border context with a stormy past, as is the case between Italy and Slovenia, the unidirectional progress of the CBRM flows cannot develop without producing some consequences, either real or perceived, or even artificially created. Indeed, the phenomenon implies a perturbation of the relations between cities and peripheries, which is a typical effect of suburbanization, but it also entails the disturbance of some sensitive balances produced by the long standing presence of the border. The sudden alteration of such balances has given birth to a growing alarmism among the Slovenian public, primarily oriented toward four main problematic issues: the environmental, landscape, social and ethno-linguistic consequences of the phenomenon.

From the environmental point of view, the Slovenian Karst region presents peculiar morphological and geological features[4] that render it an extremely fragile territory; UNESCO has included some of its areas into its World Heritage list. It is therefore clear that the high consumption of soil, produced by continuous construction interventions in the territory to accommodate the high Italian demand represents a threat to this natural heritage. Moreover, the migrants' daily cross-border commuting is increasing pollution and congestion of traffic routes, with the risk of decreasing the quality of life in the region.

The CBRM flows are also perceived to cause irreparable damage to the rural landscape, as they spread into the territory their homologation rules. The risk is that the Karst villages gradually assume the appearance of urban sprawl. One of the effects criticized is the flattening of the results of the recent building systems compared to the traditional karstal rural architecture, characterized by harmony and the use of autochthonous materials (karstal stone). The new buildings tend to look alike and present a number of elements with nothing to do with the traditional cultural landscape. The blame is placed on the lack of suitable territorial planning regulations, which paves the way to new building interventions in the form of satellite settlements, following the English New Town tradition. The territorial development is therefore influenced by the interests of major investors and real estate lobbies, whereas the local communities generally are not involved, not even when important decisions concerning territorial planning need to be taken (Lah, 2008).

This architectural flattening is not only a consequence of the new building evolutions, but also reflects the changes in the social structure of the territory and the emerging lifestyles, which underline the culture of individualism and privatism. Among the new residents, the living model created around the family is widespread: a sort of 'house-shell' where one finds shelter and develops a small, isolated domain with 'all the comforts'. Once living is reduced to residing, the suburban immigrants seem to invest little in the integration with the new social environment and prefer to maintain their reality on the other side of the border. These life models are perceived in contrast to the traditional karstal ways of life, linked to specific models of community/solidarity.

The last problematic element is partly linked to the previous point and regards the consequences produced by the CBRM flows in the cultural structure of the Karst, a territory that from the ethnical and linguistic point of view has always presented features of Slovenian homogeneity. The main problem resides in the scarce knowledge of the Slovenian language among many of the new immigrants and in the consequent difficulties in communicating with the local population in the official language of the area. This situation is perceived to have two main consequences. On the one hand, it may gradually change the sociolinguistic landscape of the karst, as the Italian language may supersede Slovenian in a growing number of communicative situations. On the other hand, it may imply an

4 These are the so called 'karst phenomena', caused by water, wind and other agents in the atmosphere, which carve the soil and form caves and underground springs.

effect of segregation, with the risk of the rise of small ethnic enclaves inserted into residential complexes, mainly independent and detached from the local social life.

These concerns have brought about the development of a series of local civil initiatives, with the mission of 'safeguarding the Karst as a national, natural and cultural heritage of Slovenia' (Civilna iniciativa Kras, 2011). In January 2011, civil initiatives presented a petition to the Slovenian government, urging it to ask the European Commission for a clause for the protection of the real estate market in the entire border area, so as to limit the sale of Slovenian real estates to 'foreign' citizens. The initiative caused a tempest with very harsh tones, replete with nationalistic shades and mutual accusations. Apart from the position assumed by the Slovenian government, which strives to avoid formal demand of the EU and would rather take preservation measures within the national legal system, the initiative represents a backward step in respect to the Europeanization process undertaken by the Italian–Slovenian borderland during recent decades.

Conclusions

From the data exposition regarding CBRM in the Italian–Slovenian border area a series of similarities with the other European case-studies can be observed, which seems to enhance the proposal to approach the phenomenon in the frame of the two theoretical perspectives discussed in the early part of the chapter. The border crossing of Italian citizens towards the Slovenian Karst is symptomatic of the advancement of suburbanization, as it presents a series of features that are strictly related to the evolution of such process: the motivations underlying residential mobility, the modalities of interaction both within the context of origin and the context of destination, the multiple effects (either real or perceived) on the receiving society. The peculiar way of life adopted by the cross-border migrants, 'suspended' between two different countries, further legitimates the study of the CBRM within the conceptual framework of transnationalism: a transnationalism that privileges the relations with the country of origin, but that also presents some elements suggesting a possible future integration of its actors, and especially of the next generations, in the new socio-cultural environment.

Once again CBRM is depicted as a fascinating, but at the same time controversial phenomenon, as it brings into the spotlight a series of contradictions, implicit in the very concept of European integration. On the one hand, it represents an interesting laboratory for bottom-up cross-border integration, as it contributes to transforming the border-barrier in a transnational space, where different people, cultures and ways of life meet and live together. On the other hand, the sudden and unexpected encounter between diversities, which were separated in the past by a border, perceived as a strong element of division and distinction, can lead to a too rapid alteration of the existing balances, to the point that tensions or conflicts, that were supposed to be laid to rest, may arise again.

The example of the CBRM is therefore symptomatic of how the efforts of the EU toward the integration of its internal borderlands may not necessarily translate into a transnationalization of the border space, but rather lead to the reiteration of national belongings and identities, thus building new 'walls' and new 'borders'. A crucial question to address in future research is whether those mobility practices are really expressions of the will to undertake a 'transnational' way of life, a living in-between of two different cultures, or, on the contrary, a simple consequence of pragmatic/utilitarian behaviours, aiming at maximizing the (economic) advantages linked to the mobility and maintaining at the same time a clear 'national' orientation towards the country of origin.

References

Ambrosini, M. (2008), *Un'altra globalizzazione. La sfida delle migrazioni transnazionali*, Bologna: Il Mulino.

Augè, M. (1992), *Non-lieux: introdution à une anthropology of supermodernity*, Paris: Seuil.

van den Berg, L., Drewett, R., Kaassen, L., Rossi, A. and Vijverberg, C.H. (1982), *Urban Europe: A Study on Growth and Decline*, Oxford: Pergamon.

Blatter, J.K. (2001), 'Debordering the World of States: Towards a Multi[h]Level System in Europe and a Multi-Polity System in North America? Insights from Border Regions', *European Journal of International Relations* 7(2): 175–209

Bohn, T. (1997), *Le mode de vie des menages Allemands de la Bande Rhenane Nord*, Strasbourg: Institut d'Etudes Politiques.

Bolzman, C. and Vial, M. (2007), *Migrants au quotidien: Les frontaliers*, Seismo: Zurich, Geneva.

Bufon, M. (2006), 'Between Social and Spatial Convergence and Divergence: an Exploration into the Political Geography of European Contact Areas', *GeoJournal* 66(4): 341–52.

Civilna iniciativa Kras (2011), *Poslanstvo*, available at civilnainiciativakras.com [accessed 30.4.2011].

Directive 2004/38/EC, *Official Journal of the European Union*, L158 of 30.4.2004: 77.

Gielis. R. (2009), 'Borders Make the Difference: Migrant Transnationalism as a Border Experience', *Tijdschrift voor Economische en Sociale Geografie* 100(5): 598–609.

Glick-Schiller, N., Basch, L. and Blanc-Szanton, C. (1992), *Towards a Transnational Perspective on Migration*, New York: New York Academy of Sciences.

Hansen, C. and Schack, M. (1997), *Grænse-pendling mellem Landesteil. Schleswig og Sønderjylland*, Aabenraa: Institut for grænseregionsforskning.

Hardi, T. (2009), 'Changing Cross-Border Movements on the Slovakian–Hungarian Border Region, after the EU accession', *Region and Regionalism* 9(1): 175–86.
van Houtum, H. and Giellis, R. (2006), 'Elastic Migration: The Case of the Dutch Short-distance Transmigrants in Belgian and German Borderlands', *Tijdschrift voor Economische en Sociale Geografie* 97(2): 195–202.
Huntington, S.P. (1996), *The Clash of Civilizations and the Remaking of World Order*, New York, London, Toronto, Sydney: Simon and Schuster.
Huijgen, G. and Reijmer, I. (2005), *Wonen in Duitsland. Een Kwalitatieve Verkenning van Actuele Ontwikkelingen*, Enschede: IandO Research.
Jagodić, D. (2009–2010), *Mobilità residenziale transfrontaliera nel contesto dell'Unione Europea: Il caso del confine italo-sloveno*, PhD dissertation in Transborder policies for daily life. Gorizia: International University Institute of European Studies.
Lah, L. (2008), 'Ključna vprašanja in odgovori o prostorskem razvoju Krasa', in Zelnik, D. (ed.) *Ohranitev kraške krajine kot razvojna priložnost Krasa*, Ljubljana: Državni svet Republike Slovenije, pp. 13–18.
Lampl, Z. (2010), 'Túl a város peremén (Esettanulmány a Pozsonyból kiköltözött felső-csallóközi és Rajka környéki lakosságról)' in Hardi, T. et al. (eds) *Magyar–szlovák agglomeráció Pozsony környékén*, Győr-Šamorín: Fórum Kisebbségkutató Intézet, pp. 77–118.
Perkmann, M. and Sum, N.L. (2002), *Globalization, Regionalization and Cross-border Regions*, Basingstoke: Palgrave Macmillan.
Portes, A., Guarnizo, L. and Landolt, P. (1999), 'The Study of Transnationalism: Pitfalls and Promise of an Emergent Research Field', *Ethnic and Racial Studies* 22: 217–37.
Prosen, A. (2008), 'Kras kot poselitveni ali kulturno-gospodarski prostor', in D. Zelnik (ed.) *Ohranitev kraške krajine kot razvojna priložnost Krasa*, Ljubljana: Državni svet Republike Slovenije, 45–49.
Ramm, M. (1999), 'Saarländer im grenznahen Lothringen. "Invasion" oder "Integration?"', *Geographische Rundschau* 51: 110–15.
Sennet, R. (1970), *The Uses of Disorder: Personal Identity and City Life*, New York: Alfred A Knopf.
Soysal, N.Y. (1994), *Limits of Citizenship. Migrants and Post-National Membership in Europe*, Chicago, London: The University of Chicago Press.
Spierings, B. and van der Velde, M. (2008), 'Shopping, Borders and Unfamiliarity: Consumer Mobility in Europe', *Tijdschrift voor Economische en Sociale Geografie* 99: 497–505.
Strüver, A. (2005), 'Spheres of Transnationalism within the European Union: On Open Doors, Thresholds and Drawbridges along the Dutch–German Border', *Journal of Ethnic and Migration Studies* 31: 323–44.
Tax Administration of the Republic of Slovenia (2011), *Number of real estates purchased by foreigners in the Republic of Slovenia. Period from 1st May 2004*

to 31st December 2010, available at http://www.durs.gov.si/si/storitve/stevilo_kupljenih_nepremicnin_v_rs_s_strani_tujcev [accessed 30.4.2011].

TendensØresund (2011), *Migration across* Öresund, available at http://www.tendensoresund.org/en/migration-across-oresund [accessed 30.4.2011].

Terlouw, K. (2008), 'The Discrepancy in PAMINA between the European Image of a Cross-border region and Cross-border Behaviour', *GeoJournal* 73: 103–16.

Velikonja, P. (2007), *Analiza razvojnih možnosti Krasa*, Ajdovščina: Občina Komen.

Vertovec, S. (2004), 'Migrant Transnationalism and Modes of Transformation', *International Migration Review* 38: 970–1001.

Zago, M. (2004), 'Il confine-ponte: la strategia', in G. Meyr and R. Pupo (eds), *Dalla cortina di ferro al confine ponte*, Civici Musei Storia ed Arte.

Chapter 11
Explaining Coverage: Why do Local Governments in Central Europe join (or not join) Euroregions?

Gergő Medve-Bálint and Sara Svensson

Introduction

Over the past two decades formalized cooperation initiatives between sub-national authorities in adjacent European countries have transformed from a relatively rare phenomenon mainly associated with a few well-known areas (Basel, the Dutch–German border and some Scandinavian examples) to a virtually omnipresent one. Including both larger working communities and micro-regions (Perkmann, 2003), there are currently about 190 such initiatives (estimation based on the records of the Assocation of European Border Regions, Perkmann 2003 and Deas and Lord, 2006).

This increased number is easily observable and, as suggested by the Europeanization literature (for instance Schimmelfenig and Sedelmeier, 2005), is probably partly due to a process of institutional transfer from the 'West' to the 'East' following the collapse of the communist regime. The interpretation of what the proliferation of cross-border cooperation stands for is contested. Do these Euroregions – here defined as formalized cooperation initiatives between sub-national authorities, often including private and non-profit actors, located close to a border in two or more countries – constitute an example of newly emerging governance structures at the local level? In a debate with certain normative overtones, some authors have argued that this is the case (Leresche and Saez, 2002), while others state that 'cross-border areas are not 'governed' in a conventional, territorial sense' (Perkmann and Sum, 2002, p. 15) or emphasize the rather marginalized role of these organizations (Kramsch, 2002; O'Dowd, 2003). At the same time, discussions around what may influence their chances for long-term viability and institutionalization have intensified. This debate is of special importance to Central and Eastern Europe (CEE), with the highest number of new, and hence vulnerable, cross-border cooperation institutions.

The goal of this chapter is to synthesize various factors suggested in the literature into a comprehensive theoretical framework and assess the relative weight of these factors when it comes to one specific aspect of institutionalization:

the ability of Euroregions to attract local governments[1] in the area where they operate. The focus is on CEE, although cautious generalizations are possible beyond this region as well. While the cohesion of a cross-border region in terms of Euroregion membership may be unproblematic in many places of Western Europe, this seems to be less so in CEE which has seen many cases of overlapping or even competing initiatives. In spite of the rich academic literature on boundaries (e.g. Rumley and Minghi, 1991; Paasi, 1999 and 2005; Kolossov, 2006; Newman, 2003; Jessop, 2002), little has been said about what defines the limits of a cross-border region in comparison with the amount of effort devoted to discussing the constructed nature of national borders. Likewise, there is an abundance of micro-level case studies of Euroregions that usually focus on 'key stakeholders' such as representatives of the organizations themselves, regional administrative and/or political bodies and major urban centres.

However, we argue that it is necessary to study a broader scope of local actors in order to gain a better insight into the mechanisms driving cross-border cooperation at the local level. While the central research question is indicated by the chapter title – why do local governments in Central Europe join or not join Euroregions? – this can be reformulated to indicate the opportunity costs: What is to be gained by joining a Euroregion as opposed to non-engagement or fostering bilateral contacts instead, such as village partnerships, school cooperation or civil society exchanges? Seeking to answer this question and also to fill in the gap in the literature in terms of available data sets, we conducted an in-depth study of a single border region in Hungary, collecting data both on local governments that are members of Euroregions and those which are not. The dataset is probably unique in that it is based on in-depth interviews with the political representatives of virtually all local governments in the region.

First, we introduce the concepts and the theoretical framework. Second, we present the methodology and the case study area within the context of the rapid emergence of Euroregions in CEE. In the subsequent section we analyse the empirical data to explain why Euroregion membership varies greatly both across different border areas (defined as regions situated along a national border), and within border regions (defined as NUTS III-level regions).

Concepts and Theoretical Framework

This study is concerned with the intersection between the organizational definition of a Euroregion and the ideational definition of a cross-border region. We acknowledge that spatial concepts, especially the concept of the region, may have multiple meanings (Hettne, 1994, pp. 136–37; Keating, 1998, p. 11; de Blij

1 Within the EU's territorial statistical system (NUTS) local governments represent the NUTS 5 or LAU 2 (Local Administrative Unit) level, i.e. the lowest level of state administration.

and Murphy, 2003, p. 17). Of special importance is the crude distinction that is made between functionalist interpretations that focus on regions as political and/or economic spaces, and constructivist interpretations that stress the importance of shared perceptions of belonging and identity. Losch's statement in his seminal article is an example of the first category. 'Impressed by the accidental way in which states are created and smashed, we are looking out for a more natural and lasting spatial order of things. [...] It is independent economic regions that we here discuss, regions not derived from but equivalent to those political, cultural, geographical units' (Losch 1938/1975, p. 97). However, Keating's description of a region as 'the result of the meeting of various concepts of space' (Keating, 1998, p. 11) rather exemplifies the constructivist understanding (see also Paasi, 2009). For the purpose of this work we adopt an instrumental approach and define a region (within a nation-state) as a NUTS 3 level territorial unit. This way we follow the functionalist tradition. Accordingly, a border area is comprised of all the NUTS 3 regions located along the border of two nation-states (see Figure 11.1).

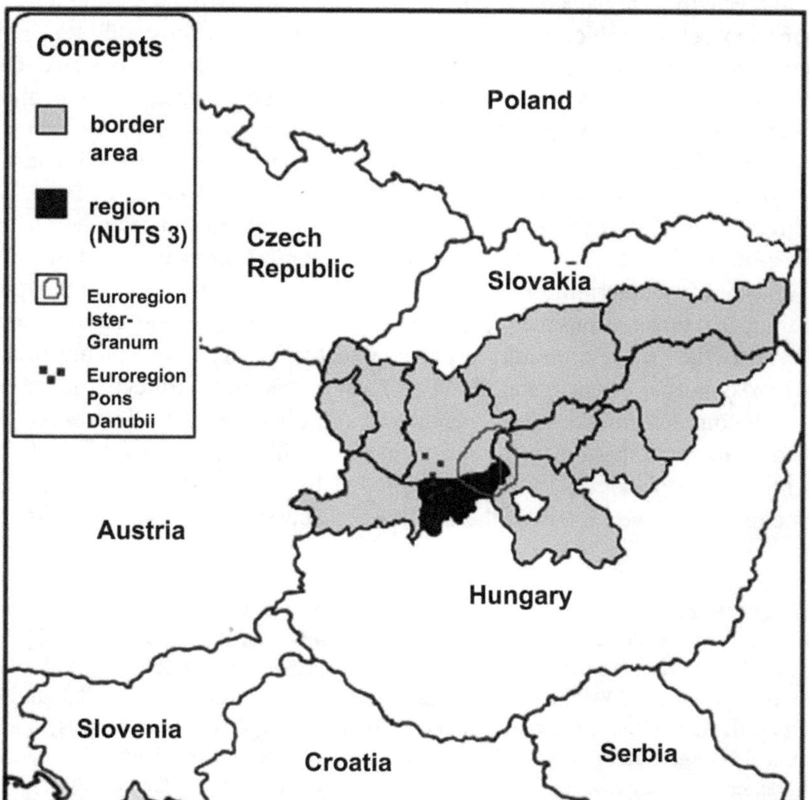

Figure 11.1 Illustration of core concepts

Further on, we refer to Euroregions as formalized cooperation initiatives between sub-national authorities (such as local or regional governments), often including private and non-profit actors, located close to a border in two or more countries. As such, a Euroregion is usually established within a single (in some cases two or more) border area and cuts across the boundaries of NUTS 3 units and state borders. Finally, moving towards more constructivist interpretations, we define a cross-border region as a socially constructed perception of a 'natural' or 'functional' region that cuts across nation state boundaries. So far our definitions are rather conventional and follow the narrative of historically coherent regions divided by unnatural national borders that is advocated by the European institutions (Celata and Coletti, 2011).

However, our departing point differs from this common narrative in (at least) two aspects. First, we do not take it for granted that all Euroregional initiatives will aspire to create such 'natural' or 'functional' cross-border regions. It is possible that an institutional framework for cooperation (i.e., a Euroregion) is set up in which the formal or de facto aspiration does not go beyond networking and/or cooperation within certain limited areas. Second, a closer look at many Euroregions would reveal that the issue of the borders and delimitations of the Euroregion itself is far from being simple. Who should be a member and who should not, is not obvious, and local governments cannot be expected to join Euroregions by default. This is especially true in those areas that have only lately opened up to organized cultural and best-practice exchanges, as well as joint policy-making for enhanced service provision and economic competitiveness. In short, the willingness of local governments to join Euroregions may show significant variation. Both these claims are substantiated by empirical findings in the subsequent sections.

We argue that there is a broad range of potential incentives and obstacles that may jointly determine membership and hence the extent of the territorial coverage that Euroregions may reach. In turn, the geographical and organizational cohesion of the Euroregions may be a crucial factor for successful and meaningful local cross-border cooperation. For this reason it is highly relevant to discuss how the complex interaction of incentives and obstacles may determine local government participation in these cooperation initiatives. A typology of such incentives and obstacles has been developed by Medve-Bálint (2012), based on a comprehensive review of the literature[2] discussing the various factors that

2 For works on cross-border cooperation in Central and Eastern Europe see especially Gabbe and von Malchus (2008), Greta (2008), Johnson (2009), Novotny (2006) and Osekowski (2000). Discussions on internal materialistic incentives can be found in Virtanen (2004) and Kennard (2004), and on internal normative incentives in Hardi (2005) and Van Houtum and Scott (2005). See also Blatter (2000) for a discussion on *polity* versus *policy* as ideational background of crossborder cooperation. There have been plenty of discussions about external incentives, mainly about EU funds, see for instance Clarke (2002), Scott (1999), Perkmann (2002), and Johnson (2009). The obstacles in the theoretical framework refer to Gorzelak (2006), Krätke (2002), Keating (2003), Perkmann (2007), and Hardi and Mezei (2003).

Incentives for local cross-border cooperation	**Instrumental**	**Normative**
Internal	need to overcome economic decline attract investments common environmental problems more efficient use of local resources	participating in European integration ("return to Europe") strong historical socio-cultural and economic ties distinct regional identity common ethnic background
External	availability of transnational funds (PHARE CBC, INTERREG) availability of national financial support established legal framework (governmental ratification of the Madrid Convention[1]; bilateral treaties)	advocacy work of transnational organizations (AEBR[2], Committee of Regions, European Commission) learning from best practices or models of other cross-border initiatives supportive policy of the central government

Figure 11.2 Incentives for local cross-border cooperation

have contributed to the emergence of Euroregions, mostly within the Central European context. The typology, which is summarized in Figure 11.2 and 11.3, constitutes the theoretical framework applied.

The structure of the tables reflect the logic of consequences and the logic of appropriateness (March and Olsen, 1989); the decision to join a Euroregion can be either instrumental (goal-oriented, utility-maximizing) or norms-based (internalizing social expectations) (Risse, 2000). Local governments may formulate their expectations towards cross-border initiatives along instrumental and normative lines: they may expect to realize material benefits arising from the cooperation or they may identify with a certain norms-based mission of cross-border cultural exchange. Besides distinguishing between normative and

Obstacles to local cross-border cooperation	Instrumental	Normative
Internal	fierce competition for resources lack of sufficient own funds lack of know-how and management skills conflicting project ideas language barriers	historical tensions, conflicts (border as symbol of identity and distinction) existing stereotypes and prejudice towards the population across the border fear of competition posed by the other side (labour market, real estate market)
External	lack of established legal framework for cross-border cooperation inappropriate external financial resources incompatible political-administrative structures lack of supportive bilateral governmental agreements	unsupportive policies of the central government conflict or tension at the governmental level between the neighbouring countries

Figure 11.3 Obstacles to local cross-border cooperation

instrumental incentives, factors that facilitate the formation of Euroregions also have a territorial aspect. There are incentives that are present or generated locally, within the border zone; while there are those that are externally created or determined as they either have a transnational origin or come from the central government. The same typology can be adapted to define the obstacles: there are instrumental as well as normative hurdles to cooperation and these may be rooted locally or have an origin that goes beyond the regional level.

Euroregions in Central and Eastern Europe and in Komárom-Esztergom

Borders usually widen the perception of distance between locations across the border thereby potentially leading to their peripheralization (Perkmann and Sum, 2002). In Central and Eastern Europe this used to be a remarkably strong mechanism before 1989, when mobility across the Iron Curtain was controlled and even discouraged, and local cross-border cooperation was limited even within the communist bloc (Kennard, 2004; Turnock, 2002). Consequently, many

Central and Eastern European border areas became economically marginalized (Böhm, 1995; Mezei, 2004; Turnock, 2002), although exceptions certainly exist. Besides causing economic decline, borders in many cases also cut through ethnic and economic entities. The frequent changes of state borders in Central and Eastern Europe in the 20th century made them contested frontiers heavily burdened with conflicts (Hardi, 2005; Van Houtum and Scott, 2005). This was further complicated by the pursuit of the principle of 'one people, one country', which caused millions to leave their homes (Eriksonas, 2006). In short, the strong presence of both incentives and obstacles to establishing institutionalized cross-border contacts in Central Europe makes this region an ideal case to study in terms of local government participation in Euroregions.

If we limit ourselves to Euroregions covering parts of the Visegrad group, which includes the four countries (the Czech Republic, Hungary, Poland and Slovakia) that inherited rather similar institutional legacies from the communist regime (Bohle and Greskovits, 2007) and that were treated as a homogenous group by transnational organizations throughout the transition, we find that already by the turn of the millennium at least one Euroregion had appeared in most of their border areas. However, as data gathered and analysed by Medve-Bálint (2012) demonstrate, participation rate of local governments show patterns of variation on two levels. First, there are differences across border areas. Second, we also find high degrees of variation within border areas at the regional level. These are the 'facts on the ground' leading up to the questions driving the research: Why do we observe variation in local government participation in Euroregions between different border areas? Furthermore, why is there cross-regional variation in membership within a single border area?

Drawing upon the data collected by Medve-Bálint (2012), it is possible to establish that external instrumental incentives in the typology developed in the previous section play a crucial role in explaining the variation between border areas. External incentives are the financial support schemes of the European Union. However, the availability and distribution of funds allocated for this purpose have been highly and persistently uneven across border areas: some of them have been consistently more preferred by the EU than others. Moreover, there seems to be an association between the amount of EU 'border' funds spent and the participation rate of local governments in Euroregions. Although this is a general tendency, there are many outlier areas that demonstrate high local government membership in Euroregions yet they did not benefit much from EU funds. This empirical fact highlights the influence of other key incentives (and obstacles) besides the impact of transnational financial resources. Moreover, this correlation at the regional level does not explain variation within a single region, making the question of why participation rates of local governments in Euroregions vary within single border areas more puzzling. This is the reason why a case-study at the regional level should clarify how these factors actually influence the decision of local governments in joining Euroregions.

Komárom-Esztergom as a Typical Case

The Hungarian NUTS 3-level county of Komárom-Esztergom was selected for our study because of its relatively high local government participation rate in Euroregions (40%) is still low enough to offer considerable variation on the dependent variable, which determines membership in Euroregions. In addition, most of the factors appearing in the analytical framework are present in the region. For instance, the large portion of ethnic Hungarians living across the border in Slovakia, and the small pockets of historical minorities (Slovaks, Germans) living in Komárom-Esztergom enable us to examine the impact of an internal normative incentive for cross-border cooperation, the common ethnicity, which is present in many Central and Eastern European border areas. The study hence displays a 'representative' or 'typical' single-case design (Yin, 2003). Finally, the case offers an abundance of data, since currently there are three Euroregions established within this territory.

Komárom-Esztergom, a rather industrial region, is situated in northwest Hungary along the Danube River, which also forms the border with Slovakia. The county is divided into 76 local governments and seven micro-regions (*kistérség*). The bordering Slovak region of Nitra has few urban centres, especially in its three micro-regions (*okres*) adjacent to Hungary, and the regional economy relies mostly on agriculture.

Before World War I Komárom-Esztergom was part of a larger Hungarian-inhabited area (the historical Esztergom County) within the territory of Great Hungary, which included the current Nitra region. Much of this zone today belongs to Slovakia. Hungarians in Komárom-Esztergom still refer to the villages and towns on the Slovak side by using their Hungarian names. It is beyond the scope of this article to discuss the recently intensified tensions between Slovakia and Hungary regarding the Slovak government's policy towards the Hungarian minority (Spectator, 2009). Nevertheless, this conflict at the government level matches with the theoretical framework's external normative obstacle to cross-border cooperation.

Within Komárom-Esztergom there are also numerous villages that preserved their Slovak or German character from the past: due to settlement policies during the Habsburg Monarchy, many villages of ethnic Slovak or German population were established centuries ago. Currently, 10 settlements have Slovak while 22 have German 'minority self-governments'. In 3 places both minorities have their self-governments established.[3] Events related to the wars of the 20th century greatly influenced the ethnic composition of many settlements. Much of the German population was expelled after WWII, and in several villages ethnic Hungarians from Slovakia were forcefully resettled as part of those plans that aimed to lower the proportion of ethnic Hungarian and Slovak minorities both in Czechoslovakia

3 Information provided by the German and Slovak National Self-governments (2010).

and in Hungary. It should be noted, however, that both the remaining German and Slovak minority in Komárom-Esztergom are largely assimilated and very few use their ethnic tongue on a daily basis whereas the ethnic Hungarians living on the Slovak side are not only much greater in numbers but have also kept their language and identity.

According to our definition, there are three Euroregions in Komárom-Esztergom: the Ister-Granum EGTC (European Grouping of Territorial Cooperation), the Hídverő/Danube Association, and the Pons Danubii EGTC. In order to examine both their membership and those local governments that do not participate in any of them, we relied on semi-structured interviews conducted with the highest political representatives of the local governments, the directly elected mayors of the 76 settlements. More than half of the interviews (43) were carried out in the premises of the local governments, and the remaining 30 were done on the phone. Although all town mayors were visited in person, for the remaining settlements we did a random selection to determine the distribution between personal and telephone interviews. In four cases high-level administrative officials received us, but the mayors subsequently approved their answers. In addition, instead of a phone interview, which was originally scheduled, we had the opportunity to meet a few mayors in person. None of this should influence the findings, as far as we can judge, based on the geographical or population characteristics of these sites. The interviews were carried out between March and August 2010. The face-to-face meetings with the mayors allowed for follow-up questions and ventures outside the immediate topic when it seemed appropriate. The interviews generally lasted for an hour, phone interviews were somewhat shorter. In addition to the interviews, we also consulted several documents regarding cross-border cooperation, which Euroregion representatives and the local governments provided for us.

In the next sections we will present the major characteristics of the three Euroregional initiatives present in Komárom-Esztergom County, keeping the focus on their membership. How did the Euroregions acquire their geographical delimitations as defined by their membership? It should be noted that two out of the three Euroregions have utilized the legal instrument EGTC (European Grouping of Territorial Cooperation). The fact that regions along Hungary's borders have been exceptionally quick in adopting this so far sparsely used instrument has been heavily promoted as a positive feature by Hungarian agencies. However, the quick spread of EGTC initiatives does not necessarily involve increased or deepened cooperation.

Ister-Granum

For more than half a century there was no permanent connection between the twin cities of Esztergom in Hungary and Štúrovo (Párkány) in Slovakia, as the bridge over the Danube – destroyed during WWII – was not rebuilt due to political reasons, fully in line with the climate of distrust described above. Finally, in 1999 the two countries reached an agreement to rebuild the bridge. It was perceived not only as the necessary precondition for setting up a regional cross-border cooperation framework, but also as an important symbol of unity. The first declaration of intent to establish local cross-border cooperation was signed in 2000 by the Slovakian Juzný micro-region and the Hungarian Esztergom-Nyergesújfalu Microregional Development Association, which altogether covered 35 local governments across the border. After these initial steps the Ister-Granum Euroregion was established in 2003 with more than 100 participating local governments from Komárom-Esztergom and the neighbouring Pest County in Hungary and from Nitra County in Slovakia. By 2003, the Euroregion covered an area of 2,200 km^2 with 220,000 inhabitants (Eck et al., 2007) – a small one from an international perspective.

Although local governments and not regions or regional associations provided the basis of membership, microregional boundaries still strongly influenced who would or would not be invited to join Ister-Granum. From Komárom-Esztergom, the Euroregion recruited local governments exclusively from the Esztergomi and the Dorogi microregions. Our interviews with the members in Komárom-Esztergom revealed two major motivations for joining. The first sees the Euroregion as a low-cost opportunity to access external funding – in the words of the mayor of Bajna, 'we thought that we had better not miss out on something'. The second is the importance of (re)uniting, at least in spirit, the Hungarian nation on a macro- as well as on a micro-level, by restoring the regional influence of Esztergom, which for centuries had been a regional centre and the seat of the historical Esztergom County before borders were redrawn after WWI. The mayor of the small settlement Sárisáp claimed that the reason for joining was 'to build relationships with the smaller settlements in our neighbour country. Those are Hungarian villages, they belonged to us in the past, and many of their inhabitants often visit us. We frequently go there as well'. The mayor of Epöl expressed a similar opinion by referring to the local government joining the Ister-Granum Euroregion as having been 'part of the historical Esztergom County'.

However, while a 'natural' cross-border region grounded in both history and function is emphasized, the vagueness of its delimitations was demonstrated when in 2009 the Euroregion took up the legal personality of an EGTC, at which stage a dozen local governments, mainly in Slovakia, opted out. Facing the choice between exit, voice and loyalty (Hirschman, 1970) when returns of the membership are perceived as too meagre, exit has been used as a strategy. However, a more common choice of local governments to voice their dissatisfaction has been to withhold membership fees (20 HUF/year/inhabitant, less than 10 Eurocents). For several years the slack in revenue was compensated by Esztergom, but recently the management of the organization – with some success – has become stricter in enforcing payment. However, powerful actors, such as the town of Nyergesújfalu, still refuse to pay the dues, referring to the lack of achievements as the main reason. These developments suggest that the geographical spread of the Euroregion may be subject to further changes, which would demonstrate the difficulty of region-building in an environment where material incentives for joining outweigh the normative ones, in spite of the rhetoric in favour of the latter.

The Hídverő/Danube Association

Although the Ister-Granum is a small Euroregion in international comparison, the Hídverő/Danube Association Euroregion is even smaller with only 60,000 inhabitants settling over 421 km^2 (Eck et al., 2007). Cooperation dates back to the early 1990s when villages located along the Danube on both sides of the border began to organize annual cultural events called 'Hídverő napok' ('Bridge building days'), but a formal cross-border organization – the Danube Euroregion – was registered only in 2003. Since then, the territorial coverage of the Euroregion has taken several turns. The formal membership was held together by Neszmély, the leading partner in the Tata Microregional Development Association and by the Slovakian '*Združenie Obcí Priatel'stva*' or '*Hídverő Társulás*' ('Bridge-Building Association') (Eck et al., 2007). Hence all local governments of the Tata micro-region in Hungary were formally members but some were significantly more active in the cooperation than others. The Euroregion was subsequently tainted by allegations of corruption towards the mayor of Neszmély, who in the end resigned in April 2008 (Neszmély, General Assembly Protocol, 2008; Népszava, 2008). The Euroregion organization was drawn into a criminal investigation and soon became defunct. However, the bonds joining the active members were not dissolved and the immediate solution was to create an organization hybrid, a combination of a Slovak-registered NGO with five Hungarian settlements being honorary members (Almásfüzitő, Dunaalmás, Kocs, Neszmély and Süttő). This is what we refer to as the Hídverő/Danube Association.

These five Hungarian members were the ones most eager to continue the cooperation, but also the ones that had partnership agreements with one of the 13 members of the Slovak association. This created a non-contiguous territorial coverage of the new Euroregion, which is also mirrored in the initiative's lack of region-building ambitions. Instead, its focus is on networking and its distinguishing feature is the extent to which the Hungarian members are loyal to its mission and working procedure. In contrast to the case of Ister-Granum, financial expectations did not feature as a reason for joining, and the overarching motivation was and still is to preserve the unity of the Hungarian people. As one of the mayors put it, 'the collaboration emerged in order to maintain Hungarian identity: the members in Slovakia are 100 per cent Hungarian settlements.' Although the initiative is based mainly on cultural exchange and has not developed into real *policy collaboration*, it does provide a forum for active *policy monitoring*. Especially the Hungarian mayors gain a close insight into the daily practices and problems of the Slovak local governments through the well-attended monthly meetings.

Pons Danubii

In 2006 discussions began on founding a third Euroregion with local membership and presence in the Komárom-Esztergom County. Towns close to the border that wished to strengthen cross-border bonds with Slovakia but at the same time did not want to succumb to the leadership of Esztergom within the Ister-Granum Euroregion, initiated the talks on the new cross-border cooperation. The Hungarian towns of Komárom, Kisbér, Tata, Oroszlány, and the Slovak towns of Kolárovo (Gúta), Nové Zámky (Érsekújvár), Komárno (Révkomárom) and Hurbanovo (Ógyalla) took part in the discussions. In the end, Nové Zámky and Oroszlány decided not to join. According to the mayor of Oroszlány, the concept was 'too general' and lacked attractive features. It is also notable that several neighbouring settlements with town status (Tatabánya, Bábolna and Ács) were left out. One of the mayors of the participating towns revealed that the inclusion of Tatabánya would have been risky since being the largest town in the county it would have dominated the Euroregion. The mayor of Tatabánya confirmed the lack of invitation.

In order to underplay the Hungarian ownership of the process, the members decided that the registration (in the form of an EGTC) would take place in Slovakia and that Komárno would serve as the administrative centre. Similarly to Ister-Granum, the new formation has a Latin name, Pons Danubii, meaning 'bridge over the Danube'.

While Pons Danubii fulfils our definition of a Euroregion (a formalized cooperation initiative between sub-national authorities in adjacent European countries), it does not form a geographically cohesive territory due to the deliberate exclusion of smaller settlements located in the neighbourhood of the six towns. In fact, it constituted a crucial issue, which local governments to include (and exclude). The outcome was not pre-determined by references to a 'natural region', and Ister-Granum actually served as a 'worst-practice example', as key actors of the new initiative considered that the high number of stakeholders in Ister-Granum had made it ungovernable.

Unlike in the case of the Ister-Granum and the Hídverő/Danube Euroregions, the motivation to reunite the Hungarian people played a less pronounced role in determining the coverage of Pons Danubii. Instead, the members focused on economic development and access to European funds. In the words of one of the mayors: 'With the EGTC we want such an organization that gives us more opportunity to deal with the important matters of the people living here without turning to the central government, but directly to Brussels instead. We already had the opportunity to apply for EU money but with the EGTC we would be able to secure even more funds.'

Cross-border Engagement of Local Governments in Komárom-Esztergom

This section presents the findings of the fieldwork in Komárom-Esztergom. We start by investigating cross-border and other international engagements outside the organizational framework of Euroregions, also including the perspective of local governments that are not Euroregion members. Then we seek to establish patterns of differences between members and non-members attempting to answer the central research question of why some local governments in Central Europe join Euroregions and others do not.

Cross-border Cooperation outside the Scope of Euroregions

Besides Euroregions, local governments can enter other forms of cross-border cooperation such as bilateral partnerships. In Komárom-Esztergom, there are 49 settlements (i.e. 67 per cent) that have established a formalized transnational partnership with at least one foreign local government and there are 9 more that have regular informal contacts with a partner abroad. Due to their bilateral nature, these transnational relationships offer more frequent and more direct contacts for the local governments than the Euroregions. Furthermore, while among the stated alternatives for joining a Euroregion was often that financial returns had been expected from the membership, bilateral partnerships have not been portrayed in this way by any of the interviewees. This may explain why most local governments were satisfied with their relationships with their partners whereas many Euroregion members expressed their discontent towards the cross-border cooperation by referring to their unfulfilled expectations.

Based on the analysis of the collected data, a number of conclusions can be drawn. The major determining factors of entering into bilateral partnerships are the size of the settlement, its distance from the border and the presence of historical ethnicity. However, the financial position of the local government seems to matter only marginally.

Settlement size is the strongest predictor of foreign partnerships: more populated local governments are more likely to have formalized bilateral partnership agreements with a foreign local government than smaller villages (Table 11.1). Out of 24 settlements with more than 2500 inhabitants, 23 have formal agreements. This is because the smaller settlements often lack those cultural associations such as folk dance groups, choirs, music bands that may be the subject of exchange and interaction within the partnership. Furthermore, they also lack the necessary infrastructure to welcome visiting groups. As the mayor of Aka with 247 residents expressed, 'we have tried to establish a bilateral partnership abroad but we did not succeed: we are too small and have nothing to offer'.

Distance from the border is another major variable that is strongly associated with the presence of foreign partnerships. Settlements located further away from the border (in terms of average travel time needed to reach the nearest border crossing by car) are less likely to engage in bilateral partnerships with foreign partners than those that are more conveniently situated to border crossing points.

In contrast, the presence of a historical minority (German or Slovak) only marginally increases the likelihood of a formal partnership agreement with a foreign settlement. From those 28 local governments in which either German or Slovak or both minorities have their own minority self-government, 22 (78 per cent) have bilateral foreign partnerships. The same figure for the 'purely' Hungarian local governments is 60 per cent (27 out of 45). Although this proportion is lower for the

Table 11.1 Settlement size and formal bilateral foreign partnerships

			Population		Total
			above 2500 inhabitants	below 2500 inhabitants	
Does the local government have partnership agreement with a foreign settlement?	no	Count	1	23	24
		% within Population	4,2%	46,9%	32,9%
	yes	Count	23	26	49
		% within Population	95,8%	53,1%	67,1%
Total		Count	24	49	73
		% within Population	100,0%	100,0%	100,0%

Pearson Chi-square: 13.355; degrees of freedom (df) = 1; p < 0.001; 0 cells have minimum expected cell count less than 5.

Table 11.2 Travel time to the nearest border crossing and formal bilateral foreign partnerships

			Travel time to the nearest border crossing			Total
			up to 25 minutes	25 – 45 minutes	more than 45 minutes	
Does the local government have partnership agreement with a foreign settlement?	No	Count	3	9	12	24
		% within Travel time	14,3%	27,3%	63,2%	32,9%
	Yes	Count	18	24	7	49
		% within Travel time	85,7%	72,7%	36,8%	67,1%
Total		Count	21	33	19	73
		% within	100,0%	100,0%	100,0%	100,0%

Pearson Chi-square: 11.653; df = 2; p < 0.05; 0 cells have expected cell count less than 5

latter group, the Chi-square test, which determines whether there is a statistically significant association between the two variables (having partnership agreements and presence of a minority), shows that the relationship is only marginally significant (p = 0.1). However, if we layer the relationship between the two variables by distance from the border, then the ethnicity factor gains statistical significance as well. Those settlements which are more remotely located from the border but have ethnic population are more likely to engage in foreign partnerships than those where the population is fully Hungarian (Table 11.3). It is also important to note that villages with German or Slovak identity tend to establish bilateral relations with German-speaking or Slovak-speaking settlements abroad. Moreover, this finding

Table 11.3 Presence of a minority and bilateral foreign partnerships layered by travel time to the nearest border crossing

				Presence of a historical minority in the settlement		Total
				no	yes (German, Slovak or both)	
Travel time to the nearest border crossing is more than 35 minutes	Presence of formal bilateral foreign partnership	no	Count	13	5	18
			% within Minority	61,9%	33,3%	50,0%
		yes	Count	8	10	18
			% within Minority	38,1%	66,7%	50,0%
	Total		Count	21	15	36
			% within Minority	100,0%	100,0%	100,0%

*Pear*son Chi-square: 2.857; df = 1; p = 0.091; 0 cells have expected cell count less than 5

suggests that the factor of cultural proximity may play a significant role in cross-border cooperation but only in conjunction with other variables.

Perhaps surprisingly, the financial position of the local governments only marginally affects their inclination to establish bilateral foreign partnerships. While 72 per cent of those local governments that assessed their financial situation as good or average had bilateral partnerships, only half of those had one that claimed to be in a poor financial position.[4] Nevertheless, according to the Chi-square test, this difference is only on the margin of being statistically significant (p =0.099). However, if the association between these two variables is layered by distance from the border, the already weak relationship disappears.

Finally, as Table 11.4 shows, having international partners seems to increase the likelihood of the local government becoming a member in a Euroregion, or it may be the other way around: being involved in a Euroregion may increase the willingness to establish cross-border bilateral partnerships.

Patterns of Differences between Members and Non-members of Euroregions

Having looked at the basis of membership in the Euroregions as well as the international and cross-border engagement of both members and non-members, we now examine if there are any differences between them and whether these differences are in line with the analytical framework outlined above.

The qualitative information provided by the mayors in combination with various descriptive statistics introduced in this section, enable us to determine which factors in the framework matter more than others. It should be emphasized though

4 The financial position of the local government was self-assessed by the mayors on a 5-point scale which was converted to the 3-point scale (good, average, poor).

Table 11.4 Membership in a Euroregion and formal bilateral foreign partnerships

			Membership in Euroregion		Total
			no	yes	
Presence of a partnership agreement with a foreign settlement	no	Count	19	5	24
		% within Membership in Euroregion	43,2%	17,2%	32,9%
	yes	Count	25	24	49
		% within Membership in Euroregion	56,8%	82,8%	67,1%
Total		Count	44	29	73
		% within Membership in Euroregion	100,0%	100,0%	100,0%

Pearson Chi-square: 5.330; df = 1; $p < 0.05$; 0 cells have expected cell count less than 5

that we did not aim to assess the exact relative weights of each factor, for which the methodology would be unsuitable. Overall, we find that the results are solid enough to justify the use of the framework and warrant some modifications for the cross-border cooperation aspect in focus, namely the motivational base of local governments to join or abstain from Euroregions. The emphasis on the micro-level is crucial, as those variables that we list here as unimportant for the micro-level may gain their significance at the meso- or macro-levels when it comes to establishing cross-border cooperation. The analysis follows the analytical grid. First, we look at the instrumental incentives and obstacles (both internal and external), and then at the normative incentives and obstacles (both internal and external).

Incentives that are instrumental and internal are derived from the common belief that the emergence of Euroregions is related to the need to resolve policy problems that cannot be dealt with if constrained by national borders. The perhaps surprising finding of our research is that these do not seem to have any importance for the decision of the local actors whether to join a Euroregion or not. Reasons such as the need to overcome economic decline, attract investments, solve environmental problems or to use local resources more efficiently rarely featured during the interviews; the exceptions were a limited number of local governments with town status.

Furthermore, as the following tables demonstrate, with respect to the perceived state of the local economy and environment, no difference can be discerned between the members and non-members. As also confirmed by the Chi-square values in tables 11.5 and 11.6, neither the financial position of the local government, nor the state of local business seems to be associated with Euroregion membership.

Table 11.5 Financial position of the local governments and membership in Euroregions

			Local government's financial position			Total
			good	average	poor	
Membership in Euroregion	No	Count	17	18	9	44
		% within Financial position	63,0%	60,0%	56,2%	60,3%
	Yes	Count	10	12	7	29
		% within Financial position	37,0%	40,0%	43,8%	39,7%
Total		Count	27	30	16	73
		% within Financial position	100,0%	100,0%	100,0%	100,0%

Pearson Chi-square: 0.191; df = 2; p = 0.909; 0 cells have expected cell count less than 5

Table 11.6 State of local business and membership in Euroregions

			State of business in the settlement			Total
			good	average	poor	
Membership in Euroregion	No	Count	10	17	17	44
		% within State of business	58,8%	70,8%	53,1%	60,3%
	yes	Count	7	7	15	29
		% within State of business	41,2%	29,2%	46,9%	39,7%
Total		Count	17	24	32	73
		% within State of business	100,0%	100,0%	100,0%	100,0%

Pearson Chi-square: 1.816; df = 2; p = 0.403; 0 cells have expected cell count less than 5

Similar conclusions were reached concerning internal instrumental obstacles (fierce competition for resources, lack of sufficient own funds, lack of know-how and management skills) as they did not make any actors decline an invitation from a Euroregion, although these issues often came up in the interviews as reasons for limiting international partnerships. However, conflicting project plans do feature as an impediment to engagement with Euroregions. For instance, discontent among the members of Ister-Granum arises from the fact that most of the initiated projects have been concentrated on the settlements located right next to the border. Tensions within Ister-Granum therefore partly revolve around different project preferences which translate into a conflict of interest between members along the border and those located further away. This may have caused the exit of those 13 local governments that chose not to take part in the EGTC.

Among the *external instrumental incentives and obstacles* the availability of EU funding does play a role: our findings suggest that the opportunity to gain access to external (EU) funds is decisive also at the micro-level. However, such instrumental (material) incentives may have ambiguous long-term consequences as these expectations may be difficult to live up to. Regarding the legal frameworks and procedures, they are usually not perceived as a problem by the local governments, although several of them mentioned that the legal registration of the Euroregions or the EGTC took longer than they had hoped for.

The normative incentives and obstacles play a different role than expected. The 'strong historical socio-cultural and economic ties' and 'common ethnic background' feature frequently as motivations ('participating in European integration' and 'distinct regional identity' much less so), and our fieldwork demonstrates that cultural or ethnic proximity plays a decisive role also in bilateral partnerships. However, we also found that mayors' attitude towards the unity of the Hungarian people across the border differs only marginally between Euroregion members and non-members. This can be exemplified by the reaction to the statement 'people speaking the same language and belonging to the same ethnic group should not be separated by borders' with which a large majority of both categories agreed (96 per cent and 88 per cent respectively). This suggests that the mayors, irrespective of the position of the local government in terms of Euroregion membership, attribute a high salience to the issue of cultural proximity, and as such potential future cross-border initiatives advocating this aspect may find a fertile ground in Komárom-Esztergom.

Internal normative obstacles (historical tensions, existing stereotypes and prejudices against the population, fear of competition) do not appear as relevant. The conflict with Slovakia is not perceived to exist on the local level, and Slovakians are not portrayed as competitors. Although the mayors are aware of the current intergovernmental conflict between the two countries, they claim that it has not trickled down to the local level and does not pose an impediment to cooperation.

At this point it is important to add that the coverage of Euroregions seems to be determined by two additional factors hitherto not included in the analytical framework. The reason why these factors did not appear in the framework was that they have been by and large neglected in the literature. The first factor is *cognitive distance*, which determines the extent to which a local government identifies with the idea of local cross-border cooperation in general and with the Euroregion in particular. Cognitive distance does not necessarily correspond to physical distance and that is why a local government may abstain from a Euroregional initiative even if it is located close to the border in the eyes of outsiders or join despite being relatively far from that. Although physical distance certainly plays a role in determining Euroregion membership, the pattern in Table 11.7 is not interpretable without accounting for cognitive distance.

Table 11.7 Distance from the border measured in travel time to the first border crossing and membership in Euroregions

			Travel time to the first border crossing point			Total
			up to 25 minutes	25 – 45 minutes	more than 45 minutes	
Membership in Euroregion	No	Count	9	18	19	46
		% within Travel time	40,9%	51,4%	100,0%	60,5%
	Yes	Count	13	17	0	30
		% within Travel time	59,1%	48,6%	,0%	39,5%
Total		Count	22	35	19	76
		% within Travel time	100,0%	100,0%	100,0%	100,0%

Pearson Chi-square: 17.147; df = 2; $p < 0.001$; 0 cells have expected cell count less than 5

The second additional factor is *administrative embeddings*, which refer to existing and past politico-administrative boundaries (such as micro-regions) that cannot be influenced by the local governments and therefore considered external factors. In our view, the reason for current administrative embeddings playing such a major role is that they provide an opportunity to resolve collective action problems. Mayors in the same micro-region have monthly meetings, which offer a setting for discussion about establishing or joining cross-border initiatives. However, for local governments outside the scope of a micro-region that collectively engages in a cross-border cooperation this factor may appear as an obstacle to join. Past administrative embeddings matter in that they influence the image of the *cross-border region*, i.e. what the Euroregion should aspire to be as is the case with the Ister-Granum Euroregion.

Figure 11.4 summarizes our main findings with respect to the theoretical framework. Factors that appear decisive in the local government's decision to join Euroregions are in bold fonts, while the factors we add to the framework appear in bold italics.

Why Do Local Governments Join Euroregions?

This chapter explored the reasons behind local governments' willingness to join cross-border cooperation initiatives in Central Europe. While the analytical framework was developed with this particular geographic area in mind, the findings probably have some relevance beyond this part of Europe.

Incentives for local cross-border cooperation	**Instrumental**	**Normative**	**Obstacles to local cross-border cooperation**	**Instrumental**	**Normative**
Internal	need to overcome economic decline attract investments common environmental problems ***converging project plans*** more efficient use of local resources	participating in European integration ("return to Europe") **strong historical socio-cultural and economic ties** distinct regional identity **common ethnic background** ***cognitive distance***	**Internal**	fierce competition for resources lack of sufficient own funds lack of know-how and management skills ***conflicting project plans*** language barriers	historical tensions, conflicts (border as symbol of identity and distinction) existing stereotypes and prejudice towards the population across the border fear of competition posed by the other side (labour market, real estate market) ***cognitive distance***
External	**availability of transnational funds (PHARE CBC, INTERREG)** availability of national financial support established legal framework (governmental ratification of the Madrid Convention; bilateral treaties) ***administrative embeddedness***	advocacy work of transnational organizations (AEBR, Committee of Regions, European Commission) learning from best practices or models of other cross-border initiatives supportive policy of the central government	**External**	lack of established legal framework for cross-border cooperation inappropriate external financial resources incompatible political-administrative structures lack of supportive bilateral governmental agreements ***administrative embeddedness***	unsupportive policies of the central government conflict or tension at the governmental level between the neighbouring countries

Figure 11.4 The modified analytical framework explaining membership in Euroregions in Central and Eastern Europe.

Based on a uniquely comprehensive set of interviews with political representatives of members and non-members of Euroregions, we conclude that the literature which constitutes the basis of the theoretical framework has underestimated the role of cognitive perceptions as well as current or previous administrative borders for the formation of Euroregions, and these two factors should therefore be added to the theoretical framework. Although cultural proximity does play a role, contrary to our expectations it determines Euroregion membership only in conjunction with other variables. Cultural proximity in this border region means common ethnicity as the majority of the population on the Slovak side of the border is of Hungarian origin. Local governments that are primarily interested in supporting the 'Hungarian cause' (i.e. the unity of the Hungarian people including Hungarians living in the neighbouring countries) are just as likely to do this via bilateral partnerships with Hungarian settlements anywhere in the Carpathian basin, not just across the border or as being a Euroregion member. We also demonstrate that the actual financial situation of the local government does not determine Euroregion membership, although politicians tend to motivate the decision to join by anticipating financial returns. Frustration with the Euroregion frequently occurs when these expectations are not fulfilled, which is the case when the main reason to join proves to be an illusionary one. This may also influence the membership composition in the long run.

In sum, the first factor determining membership as found by this study is the historical socio-cultural and ethnic ties across the border (internal normative incentive). The second is the availability of funds (an external instrumental incentive), which translates into high expectations that may, however, hamper the functioning of the Euroregions in the long run. The third is the administrative embeddedness of the local governments, in which current or historical politico-administrative institutions (external instrumental incentive/obstacle) determine the likelihood of joining. Finally, the distance to the border also matters, by which we do not only refer to physical but to cognitive distance as well.

The findings have policy relevance as they suggest that 'getting membership wrong', such as having members that joined for the 'wrong reasons' (expecting direct funding that will most likely not materialize) have serious consequences for the functioning of the Euroregion. For the same reason, bigger is not always better, when it comes to membership. Based on the findings, we may also claim that the attitudes towards cross-border cooperation are not fully in line with the agenda of major international stakeholders (such as the EU). The view that the goal of cross-border cooperation is to increase overall economic development by realizing synergy effects has not been internalized even by the representatives of cross-border cooperation organizations. Hence there remains scope for changing attitudes – if desired so – of both members and non-members.

References

Blatter, J. (2000), *Entgrenzung der Staatenwelt? Politische Institutionenbildund in grenzüberschreitenden Regionen in Europa und Nordamerika*, Baden-Baden: Nomos.

Bohle, D. and B. Greskovits (2007), 'Neoliberalism, Embedded Neoliberalism, and Neocorporatism: Towards Transnational Capitalism in Central-Eastern Europe', *West European Politics* 30(3): 443–66.

Böhm, A. (1995), 'Local Politics in Border Regions in Central Europe', *The ANNALS of the American Academy of Political and Social Science*, No. 540: 137–44.

Celata, F. and R. Coletti (2011), 'Le funzioni narrative dei confini nelle politiche di cooperazione transfrontaliera in Europa: una lettura interpretativa', *Rivista Geografica Italiana* 118(2): 219-245.

Clarke, S.E. (2002), 'Spatial Concepts and Cross-Border Governance Strategies: Comparing North American and Northern Europe Experiences', EURA Conference on Urban and Spatial European Policies, Turin

De Blij, H.J. and Murphy, A.B. (2003), *Human Geography: Culture, society and space*, 7th edn, Hoboken, NJ: Wiley.

Deas, I. and Lord, A. (2006), 'From a New Regionalism to an Unusual Regionalism? The Emergence of Non-standard Regional Spaces and Lessons for the Territorial Reorganisation of the State', *Urban Studies* 43(10): 1847–77.

Eriksonas, L. (2006), 'Historic Borders and Ethnic Arguments in Eastern Europe after 1918', in T. Lundén, (ed.), *Crossing the Border: Boundary Relations in a Changing Europe*, Eslov: Forlags AB Gondolin.

Eck, A., Jankai, N. and Ocskay, G. (2007), *A gazdasági és kulturális együttműködést megalapozó és segítő hálózat létrehozása*, Ister-Granum Euroregion Development Agency, Manuscript.

Gabbe, J. and von Malchus, V. (2008), *Cooperation between European Border Regions: Review and Perspectives*, Association of European Border Regions, Nomos: Baden-Baden.

Gorzelak, G. (2006), 'Normalizing Polish-German Relations: Cross-Border Cooperation in Regional Development', in J.W. Scott (ed.), *EU enlargement, Region Building and Shifting Borders of Inclusion and Exclusion*, Aldershot: Ashgate, pp. 195–205

Greta, M. (2008), *Transborder Cooperation, Euroregion and EU Regional Policy in the Context of Lisbon Strategy (Selected Examples of Polish Euroregions)*, Technical University of Łódz.

Hardi, T. (2005), 'Az eurorégiók, mint a határon átnyúló fejlesztés eszközei a Kárpát-medencében', available at www.magyarorszag2015.hu/tanulmanyok/Hardi_Tamas051103.pdf [accessed 3.1.2007].

Hardi, T. and Mezei, I. (2003), 'A racionalizmus és a nacionalizmus küzdelme. Eurorégiók a Magyar-szlovák határon', *Comitatus*, December, pp. 63–70.

Hettne, B. (1994), 'The regional factor in the formation of a new world order', in Y. Sakamoto (ed.), *Global Transformation: Challenges to the State system*, New York: United Nations University Press, pp. 123–66.

Hirschman, A.O. (1970), *Exit, Voice, and Loyalty: Responses to Decline in Firms, Organizations, and States*, Cambridge, MA: Harvard University Press.

Ister-Granum Euroregion, *Website* available at www.istergranum.hu [accessed 30.4.2010].

Jessop, B. (2002), 'The Political Economy of Scale', in M. Perkmann and N-L Sum, (eds), *Globalization, Regionalization and Cross-Border Regions*, New York: Palgrave-Macmillan, pp. 25–49.

Johnson, C.M. (2009), 'Cross-Border Regions and Territorial Restructuring in Central Europe: Room for More Transboundary Space', *European Urban and Regional Studies* 16(2): 177–91.

Keating, M. (1998), 'Is There a Regional Level of Government in Europe?' in P. Le Gales and C. Leusne, (eds), *Regions in Europe*, London and NY: Routledge, pp. 11–29.

Keating, M. (2003), 'Regionalization in Central and Eastern Europe: The Diffusion of a Western Model?' In Keating, M and Hughes, J. (eds), *The Regional Challenge in Central and Eastern Europe. Territorial Restructuring and European Integration*, Brussels: Presses Interuniversitaires Européennes, pp. 51–67

Kennard, A. (2004), 'Cross-border Governance at the Future Edges of the EU', in O. Kramsch and B. Hooper (eds), *Cross-border Governance in the European Union*, London and New York: Routledge, pp. 107–20.

Kolossov, V. (2006), 'Theoretical Limology: New Analytical Approaches', in T. Lundén (ed.), *Crossing the Border: Boundary Relations in a Changing Europe*, Eslov: Forlags AB Gondolin.

Kramsch, O. (2002), 'Re-imagining the Scalar Topologies of Cross-border Governance: Eu(ro)regions in the Post-colonial Present', *Space & Polity* 6(2): 169–96.

Krätke, S, (2002), 'Cross-Border Cooperation and Regional Development in the German-Polish Border Area', in M. Perkmann and N-L. Sum (eds), *Globalization, Regionalization and Cross-Border Regions*, New York: Palgrave-Macmillan, pp. 125–50.

Leresche, J.-P. and Saez, G. (2002), 'Political Frontier Regimes: Towards Cross-Border Governance?', in M. Perkmann and N-L, Sum (eds), *Globalization, Regionalization and Cross-Border Regions*, New York: Palgrave Macmillan, pp. 77–99.

Losch, A. (1975[1938]), 'The Nature of Economic Regions', in J. Friedmann and W. Alonso (eds), *Regional Policy: Readings in Theory and Applications*, Cambridge, MA: MIT Press.

March, J.G. and Olsen, J.P. (1989), *Rediscovering Institutions*, New York: Free Press.

Medve-Bálint, G. (2012), 'The Role of Euroregions in Local Development in Central Europe. Incentive Structures and Obstacles to Cooperation', in N. Bellini and U. Hilpert (eds), *Europe's Changing Geography*, London: Routledge (forthcoming).

Mezei, I. (2004), 'Rideg határok', *Comitatus*, April, pp. 64–71.

Murphy, A. (1991). 'Regions as Social Constructs. The Gap between Theory and Practice', *Progress in Human Geography* 15: 23–35.

Newman, D. (2003), 'On Borders and Power: A Theoretical Framework', *Journal of Borderland Studies* 18(1): 13–24.

Novotny, G. (2006), 'A határon átnyúló együttműködések sikerének kulcskérdései nemzetközi minták alapján', in T Kaiser (ed.), *Hidak vagy sorompók? A határon átívelő együttműködések szerepe az integrációs folyamatban*, Budapest: Új Mandátum, pp. 91–119.

O'Dowd, L. (2003), 'The Changing Significance of European Borders', in J. Anderson, L. O'Dowd and T.M. Wilson (eds), *New Borders for a Changing Europe: Cross-border Cooperation and Governance*, London: Frank Cass, pp. 13–36.

Osekowski, C. (2000), 'Barriers to Euroregional cooperation on the Polish-German border', in W. Malendowskiego and M. Szczepaniaka (eds), *Euroregiony mosty do Europy bez granic*, Warsaw: Dom Wydawniczy Elipsa, pp. 150–59.

Paasi, A. (1999), 'Boundaries as Social Processes: Territoriality in the World of Flows', in D. Newman, (ed.), *Boundaries, Territory and Postmodernity*, London: Frank Cass & Co, pp. 69–88.

Paasi, A. (2009), 'The Resurgence of the "Region" and "Regional Identity": Theoretical Perspectives and Empirical Observations on Regional Dynamics in Europe', *Review of International Studies* 35: 121–45.

Perkmann, M. (2002), 'Institutional Entrepreneurship in the European Union', in M. Perkmann and N-L Sum (eds), *Globalization, Regionalization and Cross-Border Regions*, New York: Palgrave-Macmillan, pp. 103–24

Perkmann, M. (2003), 'Cross-border Regions in Europe. Significance and Drivers of Regional Cross-border Cooperation', *European Urban and Regional Studies* 10(2): 153–71.

Perkmann, M. (2007), 'Policy Entrepreneurship and Multilevel Governance: A Comparative Study of European Cross-border Regions', *Environment and Planning C: Government and Policy* 25(6): 861–79.

Perkmann, M. and Sum, N-L. (2002), 'Globalization, Regionalization and Cross-border Regions: Scales, Discourses and Governance', in M. Perkmann and and N-L Sum (eds), *Globalization, Regionalization and Cross-Border Regions*, New York: Palgrave-Macmillan, pp. 3–21.

Risse, T. (2000), 'Let's Argue!: Communicative Action in World Politics', *International Organization* 54: 1–40.

Rumley, D. and Minghi, J.V. (eds) (1991), *The Geography of Border Landscapes*, London: Routledge.

Schimmelfennig, F. and Sedelmeier, U. (2005), 'Introduction: Conceptualizing the Europeanization of Central and Eastern Europe', in F. Schimmelfenig and U. Sedelmeier (eds), *Europeanization of Central and Eastern Europe*, Ithaca, NY: Cornell University Press, pp. 1–28

Scott, J.W. (1999), 'European and North-American Contexts for Cross-Border Regionalism' *Regional Studies* 33(7): 605–17.

SWECO (2008), *Final Report – ERDF and CF Regional Expenditure*. Prepared for the European Commission, contract number 2007. CE. 16. 0. AT. 306.

Spectator http://www.spectator.sk/articles/view/36140/2/hungary_continues_to_criticise_slovak_language_law.html [accessed 8.9.2009].

Turnock, D. (2002), 'Cross-border Cooperation: A Major Element in Regional Policy in East Central Europe', *Scottish Geographical Journal* 118: 19–40.

Van Houtum, H. and Scott, J. (2005), *Policy Considerations of Project Results: 'Good Practices' and Situational Ethics of Cross-Border Cooperation*, Berlin and Nijmegen: Exlinea Policy Paper, available at http://www.ctc.ee/exlinea/pub/EXLINEA_Policy_Paper.pdf [accessed 29.4.2010].

Virtanen, P. (2004), 'Euregios in changing Europe – Euregio Karelia and Euroregion Pomerania as examples', in O. Kramsch and B. Hooper (eds), *Cross-border Governance in the European Union*, London and New York: Routledge, pp. 121–34.

Yin, R.K. (2003). *Case Study Research. Design and Methods*, 3rd edn, Thousand Oaks, CA: SAGE, pp. 40–42.

Afterword

Olivier Thomas Kramsch

A Very Beuysian Gesture?

On 20 July, 1964, Joseph Beuys published *Lebenslauf/Werklauf (Life Course/ Work Course)*, a 'fictional' biography fusing artistic and personal aspects of his life in ways that would serve as a founding document defining his subsequent aesthetic and political persona. The first entry in the *Lebenslauf/Werklauf* register, dated 1921, reads 'Kleve Exhibition of a wound drawn together with plaster' (cited in Rosenthal, 2004, p. 159). The reference draws attention to the Northwest German border town where Beuys spent his childhood, as well as the circumstance of his birth. Further down, the register for the year 1964 reveals: 'Beuys recommends that the Berlin Wall be heightened by 5 cm (better proportions!)' (ibid.). The humour here is both personal and political. For, spiralling outwards from *Lebenslauf/Werklauf* catalogue to the many sculptures, action-performances, vitrines, multiples, phantasmagoria and environments that informed Beuys' lifelong *oeuvre*, the reader is struck by similar references to wounds, sutures, borders, divisions, fissures and frontiers of all kinds, often transposing bodily references to those indexing vaster geopolitical alignments: the artificial split between animals and humans; the tyrannical rending of gender; the existential division of the two Germanys; the destructive separation between an over-industrialized West and a 'mythical' Eurasian continent, site of putatively untainted spiritual and creative values. In each of his actions, and as foreshadowed in his Berlin Wall remark, Beuys attempted to illuminate how through art all things are conceivable, all fixed and frozen structures can be 'healed', set in motion and given life-like force so as to defuse the conditions that led to their initial freezing. This gesture would in turn be in line with Beuys admiration for a 'crazy creativity' in art whose political analogue would be revealed through his instrumental role in founding 'counter-institutional' sites such as the German Green Party, as well as parallel interventions in direct democracy ('Office for Direct Democracy through Referendum'[1]), the latter an activist organization calling for people's referenda as a way to re-democratize German society (Mesch, 2007).

The editors of *The Border Multiple* may not have had Beuysian 'crazy creativity' in mind when editing their stimulating volume of stories from Europe's contemporary borderlands, but their book title does gesture to the Beuysian legacy

1 German original: 'Buro der Organization fur Direkte Demokratie durch Volksabstimmung'.

in ways they themselves might recognize. In this Afterword we might productively enter into dialogue with this inheritance by pointing to an underlying conceptual tension at the core of the book between the abstract formalism captured by the term 'rescaling' and the unruly energies let loose by the word 'multiplicity'; explore the potentials and pitfalls of mediating between large-scale geopolitical abstractions and everyday lived experience through border-as-practice; and finally invert the analytical telescope the editors so usefully hand us so as to ponder the possibilities for a wider 'comparative turn' in European border studies.

Figure A.1 House of Hans and Franz van der Grinten, earliest benefactors of Joseph Beuys, located in the German village of Kranenburg, on the border with The Netherlands. The international political/ artistic movement that was to be named *FLUXUS*, of which Beuys was a key member (later together with George Maciunas, John Cage and Nam June Paik), was founded in the rear garage (Photo taken by author)

Re-scaling and Multiplicity: An Unresolved Tension

Andersen et al. are keen to show that in order to understand the current complexity of bordering and re-bordering practices in Europe today we must transcend the internal/external dichotomy still predominant in border studies, and grasp the process as part of a more generalized re-scaling of political space (this volume; also Keating, this volume). In my view this is a signally important argument (the first of its kind that I am aware of), in that it strives for a macro-analytical lens capable of articulating transformations within EU-space together with changes under way at its outer margins, two realms that indeed traditionally have been kept conceptually isolated from one another, notably in political science and/or critical IR studies (Larner and Walters, 2004). In order to achieve this more all-embracing view, the editors suggest developing 'new methods that are able to capture the multiplicity of the practices through which the borders are performed, as well as the way in which inside and outside are defined in time and space' (Andersen et al, this volume). To do this, they further argue, we must learn to 'avoid essentializing current trends and instead think in terms of heteronomy and diversity: the multiple borders' (Andersen et al., this volume).

These are all worthwhile moves. But if I am to play the Beuysian 'trickster' in this chapter, I would argue (as I have elsewhere, see Kramsch and Hooper, 2004) the Russian-doll house metaphor implicit in the term 're-scaling', useful for the analysis of supra- and sub-national shifts in economic and political governance *within* nation-states and *outwards* at the EU level (Keating, 1998), is inadequate for grasping the possibilities and pitfalls of extending governance capacity *across* the regional borders of nation-states. At stake is more than just a question of semantics. As has already been revealed in the course of a decade-long debate on the 'scale question' in geography and cognate disciplines, the scalar imaginary has been judged overly formalist and universalist, while tied to mechanistic understandings of state rationality almost always rooted in the particular 'spatial fixes' inherent to capitalist social relations (Harvey, 1982). It is precisely in response to the scalar arguments of these authors that many social scientists have come to recall the inherently *social* and *political* nature of all scalar formations, injecting a healthy dose of contingency in the production of all scales, from the body to the globe (Marston, 2005). The scalar imaginary draws strength from its parsimonious simplicity, but as the editors of and contributors to the *Border Multiple* are well aware, such elegant parsimony always comes at the price of chaotic, unruly, lived border experience.

It may very well be for these reasons that Andersen et al. concede that, as far as internal borders are concerned, we may need to 'multiply the notion of scale' and invent new words to describe our bordered reality, as they do when mobilizing the productive visual metaphors of 'visibility' and 'invisibility' (this volume; see in support of this view also Kramsch, 2002). I would go on to push this argument even further, and affirm that these neologisms are also supremely useful for describing dynamics at outer EU frontiers, where regimes of visibility have been at work since colonial times to render peripheral populations appropriate subjects of European imperial rule (Kramsch, 2010; Kramsch, forthcoming). Further nuancing the functionalism of any singular scalar logic, Michael Keating's intervention (this volume) trains our attention on the transformation of territory in Europe, as the latter purportedly becomes ever more 'disassociated' from the nation-state, as evidenced by the emergence of sub-national minority movements in Catalonia, the Basque country and Wales. Although his examples are drawn largely from sub-national (ie, metropolitan or sub-regional) instances rather than cross-border territories *per se*, his account has the advantage of returning our gaze to the role of political mobilization in the scalar restructuring of European space. At any rate, as with Beuys' understanding of 'doubles' in his own multifaceted *oeuvre*, the terms 'scale' and 'multiple' in this volume live in a state of unresolved tension. And perhaps it is precisely out of this tension that the stuff of border politics is made.

Rethinking the 'Monumentality' of the State: Border-as-Performance

In a somewhat rueful intervention, Thomas Wilson berates the 'borderless world' discourse so fashionable in the globalization debates of the 1990s, arguing that state borders continue to be 'the principal sites and contexts to border studies, however they are defined' (this volume; but see also earlier assertions of this kind in Paasi, 1996; Newman and Paasi, 1998). From this claim Wilson constructs a dichotomy between a field called 'border studies', putatively concerned with the 'hard' geopolitical, administrative and legal borders which continue to regulate political and economic life, and 'border theory', a diffuse academic enterprise drawing on the 'soft' intellectual eddies of cultural studies. As Wilson sees it, border theory parasitically culls border terminology from border studies in the service of its own, implicitly contrarian, agenda focusing on trans-boundary flows, cross-border identification and boundary transcendence. In using the metaphors of border studies for its own purposes, border theory thus stands charged with a free-floating idealism that denies the obdurate 'potency and relevance … of the master narratives and monumental significance of state borders in the lives of so many people of the planet' (Wilson, this volume).

But I would suggest yoking 'border theory' to the straw man of the 'borderless world' argument enacts its own kind of violence, as it forecloses the possibility of a richer dialogue between the two, something Wilson himself acknowledges in his candid concession to border studies' relatively sub-theoretical bent (ibid, this volume). In short: why must critical social theory continue to be seen as the unwelcomed guest to the tight-knit family of border studies? What is achieved – intellectually, politically – through this form of 'hard' boundary maintenance? Put provocatively: why is the Ego of border studies so 'weak' that it must circle the wagons so fiercely, inoculating itself from the effects of 'theory'? I smell disciplinary anxiety, and the loss of control, if not relevance, *vis a vis* those policymakers Wilson would have us do more truck with at the end of his essay. This is no longer a scientific problem, then, but a *political* one, placing at the forefront of our collective agendas who we write for as border scholars, and with what consequences. This question is beyond the scope of Andersen et al.'s volume, but is beginning to be explored elsewhere (Kramsch, 2012). Michael Keating's call (this volume) not to essentialize state borders as the product of some ineluctable, teleological-functionalist integration process stands as a salutary reminder that those involved in the contemporary 're-invention' of territories in Europe are often actors located far removed from the institutions of state capitols, often mobilizing those very 'culturalist' resources that Wilson implicitly perceives as epiphenomenal to the hard, underlying structures of state-centric geopolitics.

Perhaps, as with the abovementioned scale argument, and as signalled by the 'practice turn' of this volume, it may now be time to 'go beyond' the bordered/ borderless world *canard* (or at least make *paté foie gras* out of it!), as in its disciplinary effects it remains at a level of conceptual abstraction diverting urgent attention from the ways in which state borders are revealed, felt, embodied and suffered in the everyday lives of people (see Andersen, this volume). 'Monumental significance'? Indeed. But this does not mean, as the contributions to this volume richly demonstrate, that borderlanders are either borderless, de-territorialized nomads or tightly reined into an Us/Them formation by nation-state centric thinking!

'Where is element 3', Beuys noted in his diary after his 1966 action *MANRESA*, in which he enacted a split between Eastern and Western sections of the world, culminating in their eventual unification in a full cross (Adriani et al, 1979: 153–54). We may see glimmerings of such a 'synthetic' element in the contribution of Jouni Häkli, who, drawing on the ethnography and historical study of science, proposes the term 'boundary object' as a way of grasping 'any concrete or abstract element that people can use as a point of reference in their interactions' without having necessarily to reach consensus (this volume). The French-Catalan border, in his view, constitutes just such a boundary object, providing its institutional actors the 'interpretive flexibility' and 'autonomy' useful for translating more general-purpose goals into the socially rooted, multiple, socially as well as materially constructed world of the borderland. In a similar vein, Marie Sandberg (this volume) draws on Actor Network Theory (ANT) in order to suggest 'a further opening of the floor not only for actors of everyday life but for a yet broader and more analytically sturdy concept of what constitutes an actor'. This she achieves by analysing the shaping of the German–Polish border between the cities of Görlitz (Germany) and Zgorzelec (Poland). Illuminating the diverse 'modes of ordering' this particular section of the German/Polish borderland, she points to the multiplicity of practices informing cross-border integration in this historically fraught region, while offering a conceptual model which purports to privilege neither the logics of high statecraft nor the bottom-up actions of local borderlanders. As crafted by Sandberg, ANT's underlying theoretical vocabulary of 'analytical symmetry' draws us seductively into a networked world where the potentially damaging effects of structures and hierarchies are downplayed, a matter whose consequences are rather benign when it comes to analysing trans-boundary youth exchanges across the German/Polish border.

Actor Network Theory's quietism regarding questions of enduring hierarchy and power in the world system renders the geographical extension of Häkli and Sandberg's approach questionable, however, especially when applied to more (neo)colonially charged spaces such as those the EU is engaged with in its so-called North African 'Neighbourhood' (Lemberg-Pedersen, this volume). Indeed, it is precisely here, in the Southern Mediterranean spaces where Lemberg-Pedersen shows African migrants being reduced to Agambian 'bare life' through the externalization policies of EU-FRONTEX, that supposedly 'consensus-free' ad-hoc institutional arrangements and/or 'analytical symmetry' would appear to be dubious theoretical luxuries. At this juncture a return to Foucault, particularly in relation to his examination of the workings of governmental power on various forms of *subjectification*, could be a salutary move that re-trains our attention on the nationalizing, gendered, sexualized and racialized 'orderings' of difference in modernity, orderings whose spatialities cannot be merely read as the outcome of some benign 'pluralism'.

In this context, Dorte Andersen, mobilizing an ethnomethodological repertoire that is 'on the move', helpfully points to the 'political act' underlying the Janus-faced significance of all borders, demarcating a space of movement split between regulatory exclusion and radical openness. As with Marta Byrska-Szklarczyk's female 'ants' smuggling cigarettes 'Mother Courage'-like across the Polish/ Ukraine border on their very bodies (this volume), Andersen's Istrian respondents are caught between the streamlining narratives of national closure reinforced by the EU Schengen frontier, their own long-time experience of multi-lingual and multi-national belonging, and global tourism's attempt to commodify their 'cosmopolitan' identities. Between such cross-currents, they struggle to re-appropriate their subjectification in a *most* political way, as exemplified in the figure of Joska Joras, a Slovenian national who defies the expertise of Croatian cartographers to locate his house on Croatian territory. Man against map. This contest is also embodied by those for whom the nationalizing discourses of Self and Other – whether by Croatian or Slovenian elites – form part of a historical Istrian collective memory associated with the dark times of ethnic cleansing elsewhere in the former Yugoslav Republic. This raises an interesting rejoinder to Wilson's call (this volume) to 'bring history back in' to border studies, at least as it applies to Southeastern Europe. *Whose history?* And with what consequences? At least in these latter instances (see also Klatt, this volume, on the problematic historical construction of the Sonderjylland/ Schleswig cross-border region), the normative issues that might naturally find a home within the idea of 'the multiple' regain a fulsome political charge that is lacking in some of the other volume's contributions. In sum, and paraphrasing Beuys, we might say: '*Jeder ist eine (Grenz)kuenstler!*[2]', but not always under conditions of her own choosing.

Reading productively across these interventions shows, once again, that the use of critical social theory can illuminate core problematics within the more geopolitically-orientated field of 'border studies' proper, with far-reaching effects. And it is by way of such felicitous theoretical conversations that we might open onto a discussion exploring the wider comparative vocation of European border studies, a matter glanced at by the editors of this volume.

Inverting the Telescope: Towards a 'New Comparativism' in European Border Studies?

By focusing on the multiplicity of practices informing the production of borders in Europe today, the volume's editors are keen to avoid reifying an inside/outside dichotomy to the contemporary transformation of European space. Such essentializing would posit a pure theatre of movement within Schengenland, and an impenetrable 'Fortress Europe' on the EU's outer perimeter (Andersen et al, this volume). What such a conceptual frame would miss, in their view, is a significant 'continuity of European border regimes and the complexity of bordering processes when performed in practice' (ibid: 7). The call for greater analytical attention to the continuities in border practices linking internal and external border regimes would thereby converge with the editors' broader goal of connecting these practices 'with more general economic and political changes occurring not only internally to Europe proper but at a global scale' (ibid:7).

Irrespective of whether this aim has ultimately been achieved or not within the pages of the volume, at stake here, I believe, is an important attempt within border studies to draw us away from an understanding of borders and border regions as merely local arenas marked by idiographic uniqueness (as suggested as recently as 2002 by Kaplan and Häkli), towards seeing them as part and parcel of a transnationally relational, and global dynamic (see Rumford, 2010). In this context, the recourse of the editors to Benedict Anderson's famous 'inverted telescope' is not haphazard, as it points to the need for a more open, critically comparative perspective capable grasping internal and external European bordering processes within a shared analytical 'lens' (for recent moves in this direction within critical urban and regional studies, see Robinson, 2011).

To understand more fully what such a perspective might offer the study of European borders, it is worth quoting at length the relevant passage from Anderson's *oeuvre*:

2 Beuys' original formulation was 'Jeder ist ein Kuenstler' (ie, 'Everyone is an artist'), a motto which expressed his belief in the inherently democratic nature of art

> On February 2, 1963, about a year after my initial encounter with what I had been trained to imagine as 'Southeast Asia', I had a strange experience to which at that time I could not give a name. The then President of Indonesia, Sukarno, was to receive an honorary degree from the University of Indonesia, and he had invited the *corps diplomatique* to join the students and faculty for the occasion … All went pleasantly until, out of the blue, he began to talk about Adolf Hitler, and in a strange manner—not as a mass murderer, not even as a fascist and anti-Semite, but as a nationalist … I felt a kind of vertigo. For the first time in my young life I had been invited to see my Europe as through an inverted telescope. Sukarno regarded himself as a man of the left, and he was perfectly aware of the horrors of Hitler's rule. But he seemed to regard these horrors with the kind of calm with which a devout Christian contemplates the centuries of massacres and tortures committed in His Name—or perhaps with the brisk distance from which my school teachers had spoken of Genghis Khan, the Inquisition, Nero, or Pizarro. It was going to be difficult from now on to think of 'my' Hitler in the old way. I did not have a good name for this experience till almost a quarter of a century later, when I was in the Philippines and teaching myself to read Spanish by stumbling through Jose′ Rizal's extraordinary nationalist novel *Noli Me Tangere*. There is a dizzying moment early in the narrative when the young mestizo hero, recently returned to the colonial Manila of the 1880s from a long sojourn in Europe, looks out of his carriage window at the municipal botanical gardens, and finds that he too is, so to speak, at the end of an inverted telescope. These gardens are shadowed automatically—Rizal says *maquinalmente*—and inescapably by images of their sister gardens in Europe. He can no longer matter-of-factly experience them, but sees them simultaneously close up and from afar. The novelist arrestingly names the agent of this incurable doubled vision *el demonio de las comparaciones* … So that's what it was in 1963, I said to myself: the spectre of comparisons. (Anderson, 1998, pp. 1–2)

Ben Anderson's '*demonios*' appear all too fleetingly in this volume, and that is a pity, as they might have provided a compelling metaphor (and method) for connecting internal European bordering processes to properly global space. This could be achieved by way of a postcolonial vision which extends beyond a-historical, overly mechanistic and state-centric scalarities internal to EU-space to embrace an *art of seeing* the EU from the vantage point of Europe's ex-colonial frontier peripheries, now increasingly occupying centre stage in world geopolitics. These frontiers, as Maier (2006) and Balibar (2009) remind us, have historically played an important recursive role in the shaping of internal, national- European borderlands, serving as the arenas where national archetypes of bourgeois citizenship, liberal economy and governance were fashioned and reimported into the metropoles (see also Cooper and Stoler, 1997). Closer to our disciplinary home, the 'vertigo' experienced by Anderson upon listening to 'Sukarno's Hitler' might equally be experienced by the scholar of Belgian borders who stumbles upon the formative role of Congolese decolonization in the accentuation of the Walloon/

Flemish linguistic divide (Mandel, 1963; Sidaway, 2002); the role of the Algerian War in the delimitation of the hexagonal French Vth Republic (Ross, 1988); or the current use of the EU's Interreg 'manual' for transboundary cooperation as a model in West Africa (Kramsch and Brambilla, 2007). *Demonios de las comparaciones*, all. But for these cross-border demons to have been mobilized appropriately in all their effulgent multiplicit(ies), the editors may well have had to organize another type of conference, taking place in a different, perhaps more appropriately non-European setting. In any event, this wider, critical comparative frame for the study of European borders and frontiers remains the task of our generation. We can only look forward to that adventure, and remain grateful that the editors have flagged this challenging agenda for the future of our field.

The comparative study of borders *within* the European Union, of course, has a long and rich pedigree. Thomas Wilson recounts how after preparing a conference on just this thematic relating to Northern Ireland in the early 1990s, he and co-editor Hastings Donnan laughed to themselves at the thought that 'if we had arrived at borders as a research topic and arena for comparative anthropological investigation, then perhaps all that had been interesting on the subject had already been said!' (Wilson, Chapter 4 this volume). Some 20 years on from Wilson and Donnan's humorous outburst, we may now retrieve it as an invitation to continue in our wise laughter, knowing well that from the vantage point of *Café Europa* (Figure A.2), we all have yet to say the last word.

Figure A.2 Having a 'Wilsonian' laugh at *Café Europa*? German café, outskirts of Kranenburg, frequented mostly by Dutch patrons, en route to Dutch/German border (Photo taken by author, with intellectual debts to Dorte Andersen)

References

Anderson, B. (1998), *The Spectre of Comparisons: Nationalism, Southeast Asia and the World*, London: Verso.

Adriani, G., Konnertz, W. and Thomas, K. (1979), *Joseph Beuys*, Cologne: Du Mont.

Balibar, E. (2009), 'Europe as Borderland', *Environment and Planning D: Society and Space* 27(2): 190–215.

Cooper, F. and Stoler, A.L. (eds) (1997), *Tensions of Empire: Colonial Cultures in a Bourgeois World*, Berkeley: University of California Press.

Harvey, D. (1982), *The Limits to Capital,*. Oxford: Basil Blackwell.

Kaplan, D.H. and Häkli, J. (eds) (2002), *Boundaries and Place: European Borderlands in Geographical Context*, Lanham, MD: Rowman & Littlefield.

Keating, M. (1998), *The New Regionalism in Western Europe: Territorial Restructuring and Political Change*, Cheltenham, UK; Northampton, MA: Edward Elgar.

Kramsch, O. (2002), 'Re-imagining the Scalar Topologies of Cross-Border Governance: Eu(ro)regions in the Post-Colonial Present', *Space & Polity* 6(2): 169–96.

Kramsch, O. (2010) 'Camuspace: Towards a Geneaology of Europe's De-Colonial Frontier', in C. Brambilla and B. Riccio (eds), *Transnational Migrations, Cosmopolitanism and Dis-Located Borders*, Rimini: Guaraldi, pp. 87–118.

Kramsch, O. (forthcoming) '"Swarming" at the frontiers of France, 1870–1885', in T. Wilson and H. Donnan (eds), *The Blackwell Companion to Border Studies*, Malden, MA: Blackwell.

Kramsch, O. (ed.) (2012), Dossier: '*Frontiere Phalanstere*? Crossing the borders between "theory" and "activism"', *ACME: An International E-Journal for Critical Geographies*.

Kramsch, O. and Brambilla, C. (2007), 'Transboundary Europe through a West African Looking Glass: Cross-Border Integration, "Colonial Difference" and the Chance for "Border Thinking"', *COMPARATIV* 17(4): 95–115.

Kramsch, O. and Hooper, B. (2004), *Cross-border Governance in the European Union*, London and New York: Routledge.

Maier, C.S. (2006), *Among Empires: American Ascendancy and its Predecessors*. Cambridge, MA: Harvard University Press.

Larner, W. and Walters, W. (eds), *Global Governmentality: Governing International Spaces*, London: Routledge.

Mandel, E. (1963), 'The Dialectic of Class and Region in Belgium', *New Left Review* 20: 5–31.

Marston, S.A., Jones, J.P. III and Woodward, K. (2005), 'Human Geography without Scale,' *Transactions of the Institute of British Geographers* 30: 416–32.

Mesch, C. (2007), 'Institutionalizing Social Sculpture: Beuys' Office for Direct Democracy through Referendum Installation (1972)', in C. Mesch and V.

Michely (eds), *Joseph Beuys: The Reader*, London and New York: I.B. Taurus, pp. 198–217.

Newman, D. and Paasi, A. (1998), 'Fences and Neighbours in the Postmodern World: Boundary Narratives in Political Geography', *Progress in Human Geography* 22(2): 186–207.

Oeller, A. (1997), 'Fluxus at the Border: Aachen, July 20, 1964', in E. Gillen (ed.), *Deutschland Bilder: Kunst aus einem geteilten Land*, Cologne: Dumont, pp. 47-62.

Paasi, A. (1996), *Territories, Boundaries and Consciousness: The Changing Geographies of the Finnish-Russian Border*. Chichester: John Wiley & Sons.

Robinson, J. (2011), 'Cities in a World of Cities: The Comparative Gesture', *International Journal of Urban and Regional Research* 35(1): 1–23.

Rosenthal, M. (2004), 'Joseph Beuys: Staging Sculpture', in M. Rosenthal, S. Rainbird and C. Schmucki (eds), *Joseph Beuys: Actions, Vitrines, Environments*, The Menil Collection, Houston, TX: Yale University press, pp. 10–135.

Ross, K. (1988), *The Emergence of Social Space: Rimbaud and the Paris Commune*. Minneapolis: University of Minnesota Press.

Rumford, C. (2010), 'Global Borders: An Introduction to the Special Issue', *Environment and Planning D: Society and Space* 28(6): 951–56.

Sidaway, J. (2002), 'Guest Editorial: Europe and the Spectre of Comparisons', *Space and Polity* 6(2): 133–9.

Index

FSC
www.fsc.org
MIX
Papier | Fördert gute Waldnutzung
FSC® C083411

Kolibri 360 GmbH
Ferdinand-Jühlke-Straße 7
99095 Erfurt, Deutschland
produktsicherheit@kolibri360.de